Adapting to a Changing Environment

Adapting to a Changing Environment

CONFRONTING THE CONSEQUENCES OF CLIMATE CHANGE

Tim R. McClanahan
and
Joshua E. Cinner

OXFORD
UNIVERSITY PRESS

Oxford University Press, Inc., publishes works that further
Oxford University's objective of excellence
in research, scholarship, and education.

Oxford New York
Auckland Cape Town Dar es Salaam Hong Kong Karachi
Kuala Lumpur Madrid Melbourne Mexico City Nairobi
New Delhi Shanghai Taipei Toronto

With offices in
Argentina Austria Brazil Chile Czech Republic France Greece
Guatemala Hungary Italy Japan Poland Portugal Singapore
South Korea Switzerland Thailand Turkey Ukraine Vietnam

Published by Oxford University Press, Inc.
198 Madison Avenue, New York, New York 10016
www.oup.com

Library of Congress Cataloging-in-Publication Data

McClanahan, T. R.
Adapting to a changing environment : confronting the consequences of
climate change / Tim R. McClanahan and Joshua E. Cinner.
p. cm.
Includes bibliographical references and index.
ISBN 978-0-19-975448-9 (hardcover : alk. paper)
1. Climatic changes. 2. Global environmental change. I. Cinner,
Joshua E., 1972- II. Title.
QC903.M395 2011
304.2'5—dc22
2011007320

1 3 5 7 9 8 6 4 2
Printed in the United States of America
on acid-free paper

CONTENTS

Preface vii
Acknowledgments ix
List of Abbreviations xi

1. Climate Change, Resources, and Human Adaptation 1

2. The Global Context of Marine Fisheries 7

3. Climate Change and Oceanography 22

4. Climate Change and the Resilience of Coral Reefs 37

5. Climate Change and Coral Reef Fishes and Fisheries 57

6. Vulnerability of Coastal Communities 67

7. Coastal Communities' Responses to Disturbance 84

8. Linking Social, Ecological, and Environmental Systems 100

9. Building Adaptive Capacity 115

10. Managing Ecosystems for Change 134

11. Confronting the Consequences of Climate Change 150

Bibliography 154
Index 181

PREFACE

The changing climate may fundamentally alter the land and sea as we know it. For those who depend on the beauty and bounty of the Earth's natural resources for their livelihoods—especially the world's poor—these changes could spell disaster. The problems climate change poses are complex, as are the ways in which societies cope with and adapt to change. Understanding and addressing these problems requires bridging diverse fields within the geophysical, ecological, and social sciences.

An ecologist and a social scientist, we have spent the last decade working together to integrate these fields. We approach the book from the perspective that social and ecological systems are intimately linked. Social processes, which can include cultural, political, and economic characteristics of society, influence the ways that people use and manage natural resources. Likewise, ecological conditions and processes can influence the societies' well-being.

Using this interdisciplinary approach, this book synthesizes, in simple terms, the rapidly emerging fields of climate change science and human adaptation and develops a practical framework for much-needed policy and adaptive responses. The framework addresses the differential responses of the environment, ecology, and people in affected areas, and identifies the policy action priorities based on this heterogeneity. We hope that this type of integrated analysis and problem solving will lead to policy actions that promote appropriate and lasting adaptations.

As a focal lens for these integrated climate change issues, we explore coral reefs and the coastal societies that depend on them throughout the eastern coastline of Africa and the islands of the western Indian Ocean. This is where many of the Earth's most impoverished people live. Here, both ecosystems and peoples' livelihoods are extremely sensitive to climate disturbances. Monsoonal rains, which are heavily influenced by climatic patterns, provide nearly all of the rainfall for the region's agriculture. Likewise, the islands and coasts are fringed by coral reefs, which provide livelihoods for millions of fishers and their dependants in the region, but are one of the most climate-sensitive ecosystems. Considerable climate impacts have already occurred to the regions coral reefs-and even more severe ones are expected. This region, like others in poor tropical countries, has neither contributed much to rising greenhouse gas emissions, nor is it likely to contribute greatly to the efforts to mitigate climate change. Countries in the region will have little choice but to adapt, but these efforts will face considerable challenges from persistent poverty, implementing decisions, corruption, and other prevalent socioeconomic conditions.

The challenges of undertaking climate science, making the findings accessible, and catalyzing action are considerable, but this region is where these efforts and responses are most needed. Harsh realities will need to be confronted with decisions that increase the chances for successful adaptation. Although our book focuses on a specific geographic region and ecosystem, the conceptual framework we develop is applicable to most regions and climate change problems. Those interested in how climate change may influence other regions or systems can adapt the framework and approach we develop beyond the specific case we present.

ACKNOWLEDGMENTS

We thank the Western Indian Ocean Marine Science Association (WIOMSA) through support from the Swedish International Development Corporation Agency (SIDA), and the John D. and Catherine T. MacArthur Foundation for their support of this book. WIOMSA's vision of interdisciplinary research toward practical problem solving has inspired much of our most satisfying research, and this book represents an effort to contextualize, synthesize, and build on these ideas. We also thank a large number of colleagues who have helped us during our research and preparation of this book, including helpful reviews and comments. These include Caroline Abunge, Eddie Allison, Mebrahtu Ateweberhan, Andrew Baird, Andrew C. Baker, Duan Biggs, Lionel Bigot, Tom Brewer, Katrina Brown, Henrich Bruggeman, Pascale Chabanet, Dee Cinner, Susan Clark, Tim Daw, David Feary, Chris Funk, Rene Galzin, Kajsa Garpe, Nick Graham, Mirielle Guilliame, Christina Hicks, Cindy Huchery, Simon Jennings, Albogast T. Kamukuru, Yves Letourneur, U. Lindhal, Joseph Maina, Nadine Marshall, Caleb McClennen, Nyawira Muthiga, Mohammed Sulieman, Ruby Moothien Pillay, Chris Muhando, Marcus Ohman, Nick Polunin, Carlos Ruiz Sebastián, Charles Sheppard, Mark Spalding, Selina Stead, Rashid Sumaila, Saleh Yahya, Andrew Wamukoto, Shaun Wilson, and Jens Zinke.

LIST OF ABBREVIATIONS

AHP	Analytic Hierarchy Process
BMU	beach management units
BP	before the present
CV	coefficient of variation
EEZ	exclusive economic zone
ENSO	El Niño southern oscillation
FAO	Food and Agriculture Organization
GDP	gross domestic product
GELOSE	Gestion Locale Sécurisée
HDI	human development index
IOD	Indian Ocean dipole
IPCC	Intergovernmental Panel on Climate Change
ITCZ	inter-tropical convergence zone
LMMA	Locally Managed Marine Areas network
MPA	marine protected area
MMSY	multi-species maximum sustainable yield
NGO	nongovernmental organization
NOAA	National Oceanographic and Atmospheric Administration
OECD	Organization for Economic Co-operation and Development
PAR	photosynthetically active radiation
PDO	Pacific decadal oscillation
PPP	purchasing power parity
SIDA	Swedish International Development Corporation Agency
SST	sea-surface temperature
UNCLOS	United Nations Convention on the Law of the Sea
UV	ultraviolet
WIO	western Indian Ocean
WIOMSA	Western Indian Ocean Marine Science Association
WWF	World Wildlife Fund

1

Climate Change, Resources, and Human Adaptation

Human relationships with nature can follow different paths. Sometimes the path leads to the collapse of both ecosystems and society. History shows that the directions down this path are simple—unsustainable practices lead to severe environmental damage. This environmental damage has various harmful feedbacks into society. The resultant struggle for natural resources can lead to starvation, ethnic tension, wars, increased diseases, emigration, and turmoil among governments. Importantly, these struggles are not restricted to ancient history but, rather, have occurred recently in many places where resources and politics interact, including Rwanda, Ethiopia, Eritrea, Zimbabwe, Kenya's Rift Valley and coast, and Sudan in the past two decades.

History also shows that these environmental damages are often amplified by changes to the climate, such as acute warming, drought, or floods. Sometimes when overexploitation of resources is combined with even small changes to the climate, thresholds are crossed that lead to surprising changes in the ecosystems that people depend on for food. Quite simply, this path may look fine and even be very profitable for a long time, but even a small change in climate can lead to disastrous situations that are difficult to reverse.

This path is not inevitable and it is possible to successfully confront the consequences of climate change. Societies can avoid these harmful consequences by recognizing that resource use and climatic change are real problems that are influenced by human actions and by organizing such that these dangerous thresholds are not crossed. This alternate path is not an easy one, for it requires significant investments that will help both society and ecosystems cope with change.

The purpose of this book is to outline some challenges and opportunities toward this alternate path. As a lens to explore this issue, we focus our attention on an understudied area of the western Indian Ocean (WIO), where the issues of natural resource abuse and poverty desperately need solutions and where climate change and coastal resources are a critical component of people's present and future. Here, climate change is not some distant possibility far into the future. It has already happened and is likely to get much worse.

Already, for example, unprecedented increases in ocean temperatures have caused as much as 90% of the corals to die in some parts of the WIO in 1998 (Ateweberhan and McClanahan 2010).

It is these coral reefs, their intimate relationship with coastal societies, and the ways that both of these will be affected by climate change that will be our main focus. Together, we have over 45 years of experience studying coral reefs and the coastal societies that depend on them. This book represents our effort to bridge the often-disparate disciplines of social science and ecology to address the complicated and inter-related problems of climate change, resource abuse, and poverty. The issues we raise and frameworks we develop to address them are not limited to our focal case study in the WIO. They are applicable to a wide range of social and ecological systems, and we hope they inspire scientists, managers, policy makers, students, and others who are concerned about the climate challenge.

The Climate Challenge

Climate change is one of the greatest challenges society faces. The Intergovernmental Panel on Climate Change (IPCC) is the leading body for the assessment of climate change and its potential consequences to society and ecosystems. Its general conclusions are that the planet is warming at 0.1–0.2°C per decade as a result of increased greenhouse gasses that trap the Earth's heat. This is making the climate more variable and, on average, wet areas will become wetter and dry areas drier (Allan and Soden 2008).

These IPCC predictions are based on the confluence of scientific theory, climate models, and time series of various environmental measurements that spanned at least 20 years (Parry et al. 2007). An impressive total of around 29,000 time series have been examined and 90% of them support the climate warming scenarios. For marine and freshwater in general (86 time series) and Africa (7 time series) and western Indian Ocean in particular, the number of time series are sparse and hardly a strong basis for detecting climate change and making changes in policy (Richardson and Poloczanska 2008). Furthermore, climate models for Africa and the Indian Ocean suffer a number of problems that reduce the accuracy of specific climate change predictions made from coarse-scale models (Funk et al. 2008). The specifics of these problems will be discussed further in Chapter 3.

As will become apparent in this book, these modeling problems do not mean there is no evidence for climate change in the WIO region—simply that the evidence is currently sparse according to the IPCC criteria. The IPCC report offers this as the explanation for the imbalance: "lack of access by IPCC authors, lack of data, research and published studies, lack of knowledge of system sensitivity, differing system responses to climate variables, lag effects in responses, resilience in systems, and the presence of adaptation." One of the purposes of this book is, therefore, to review the evidence for climate change among the region's coasts and fill in some of these gaps in knowledge. We take an in-depth look at the climate over the past several thousand years and the mechanisms whereby climate change affects both coral reefs and the critical fisheries they support.

The Impacts of Climate Change on Society

The IPCC reports that Africans are likely to be severely impacted by climate variability and change (Boko et al. 2007). For example, by 2020, between 75 million and 250 million people in Africa will be exposed to increased water stress due to climate change. Climate change is a critical issue in Africa and particularly along the WIO coast because it is expected to have large influences on food production in what is considered one of the more food insecure parts of the world. Food production is likely to be reduced through (1) coral bleaching caused by ocean warming events that will affect reef fish targeted by fishers (Graham et al. 2007, Cinner et al. 2009a); (2) rising water temperatures that will reduce fisheries resources in large lakes (Boko et al. 2007); (3) changes in monsoon conditions and yields from rain-fed agriculture that could be reduced by up to 50% by 2020 because shifting, irregular, and declining rainfall along the African coastline will reduce the total and per capita production (Boko et al. 2007, Bowden and Semazzi 2007, Funk et al. 2008); (4) high intensity storms, such as cyclones, that are likely to increase in frequency and shift in distribution and have localized but considerable impacts on food production; and (5) coastal areas are subject to sea-level rise, which may cause saltwater intrusion in coastal agricultural areas.

The impacts of such trends are likely to be amplified by pervasive social conditions in the region, such as extremely high population growth, paralyzing poverty, heavy exploitation of natural resources, and weak governance. In addition, the spread of malaria may increase with rising temperatures and along with high rates of HIV/AIDS (human immunodeficiency virus/acquired immunodeficiency syndrome) may further undermine the capacity of societies to cope with or adapt to change (Hay et al. 2002). Fragile gains in Africa's development over the past few decades may, therefore, be severely compromised by the increasing threat of climate change.

The convergence of these forces is cause for concern but not resignation. The way that societies in this region organize around these social, environmental, and ecological changes will prove critical to reducing human suffering and sustaining ecosystems (Turner et al. 2010). In this book, we examine how vulnerable societies in the WIO are to the impacts of climate change and also highlight potential responses to this change.

Confronting Change: Adaptation, Resilience, and Vulnerability

The ways that people and ecosystems are affected by and deal with change are influenced by the three inter-related concepts of adaptation, resilience, and vulnerability. These concepts are central to this book and require some introduction because they are used by the general public but also have different meanings in various academic disciplines.

ADAPTATION

Adaptation, as commonly used by geographers and the general public, includes a mixture of responses that can diminish risks. In the context of climate change, distinctions are made between adaptation and mitigation. Adaptation refers to actions undertaken to reduce the effects of climate stresses on human and natural systems. Examples of adaptation measures include the construction of sea walls to manage sea-level rise and alteration of farming or fishing practices. Mitigation refers to attempts to reduce the magnitude of our contribution to climate change, such as reducing greenhouse gas emissions or increasing carbon storage by planting more trees. This book will focus on adaptation, primarily at the individual and local community level, and does not address the topic of mitigation, which is covered in detail in a number of other works.

Biologists have a very different meaning for adaptation, considering it an evolutionary process whereby a population becomes more successful or better adapted in its environment. Biologists distinguish between adaptation and acclimatization. Acclimatization is phenotypic, meaning the way that individual organisms adjust their physiology to a changing environment (Edmunds and Gates 2008). Adaptation is genotypic connoting changes in the frequency of genes in a population over time through their differential growth and persistence in the broader population. Acclimatization is rapid and adaptation is slower. The book will use adaptation in the way that geographers and the general public use it and, when genetic change is inferred, it will be referred to as genetic adaptation.

RESILIENCE

Resilience is frequently defined as the amount of external social, political, or environmental disturbances a system can to cope with or adapt to before switching to an alternative state (Folke et al. 2002, Marshall and Marshall 2007). A highly resilient system will be able to resist change or quickly recover from it. Resilience can be a positive attribute (a desirable system can absorb multiple perturbations before shifting to an undesirable state) or a negative one (an undesirable system state may be persistent despite multiple attempts to change it to a desirable one). Of course, what some consider a desirable environment may be undesirable to others, but a desirable ecosystem is generally one that provides important ecological services to people, such as fish or protection of the shoreline, in the case of coral reefs.

There are several important concepts espoused by the people who think and write about resilience that are central to this book. First is the concept that social and ecological systems are linked, in what some scientists call a coupled or linked social-ecological system. The idea behind linked social-ecological systems is that human actions and social structures, including economic and political organization, profoundly influence environmental, human population, and natural ecological dynamics, and vice versa (Hughes 1994, Adger 2006). Research that ignores these linkages is likely to miss critical feedbacks between social and ecological dynamics that may lead the social-ecological system toward a situation undesirable for people that may be difficult or impossible to reverse (Holling et al. 1998). An example of these types of undesirable situations is a reinforcing cycle of poverty and resource misuse known as a poverty trap (discussed in Chapter 7).

Second, social-ecological systems regularly undergo changes or perturbations. Systems with a history of disturbance may have lost their more vulnerable components, which may play a critical role in maintaining important ecological processes This loss of resilience may be difficult to notice; on the surface, things appear unchanged, but then small changes can lead to a dramatic shift in the social-ecological system, in what is known as an ecological "phase shift" (Hughes 1994). Thus, defining the thresholds where social-ecological systems undergo dramatic changes is a critical priority, and resource managers need to ensure that social-ecological systems do not get pushed too close to these thresholds. We describe where key thresholds may lie in coral reef social-ecological systems and what can be done to buffer them in Chapters 5 and 10.

VULNERABILITY

Vulnerability is often perceived as the opposite of resilience. It is the susceptibility to harm, and the term is most often applied to people or social-ecological systems. Vulnerability can be influenced by local dependence on natural resources, poverty, corruption, and a range of other factors that are described in detail in Chapter 6. Vulnerability is often perceived as having three distinct components: (1) *exposure*, which is the degree to which a system is stressed by the environment. This can be characterized by the magnitude, frequency, duration, and spatial extent of a climatic event, such as strong climatic oscillation or a cyclone; (2) *sensitivity*, which is the degree to which stress actually modifies the response of a system and is frequently affected by things such as local level dependence on marine resources; (3) *adaptive capacity*, which refers to the conditions that enable people to adapt to change. People with high adaptive capacity will likely be able to adapt to shifts brought about by climate change and take advantage of opportunities they create. These components, particularly exposure and adaptive capacity, are central to our exploration of both the problems and potential solutions to the climate challenge.

A Framework for Confronting the Climate Challenge

To investigate potential solutions to the climate challenge, we develop a general framework that is applicable to a wide range of social-ecological systems, issues, and scales (Fig. 1.1). The framework simultaneously considers three components that are critical to natural resource use and management dilemmas in a climate change context: (1) the exposure to climate change; (2) the capacity of society to adapt to change or adaptive capacity; and (3) ecological conditions of key resources, such as fish and their habitats.

Every social-ecological site, such as a fishing village, will have a variety of measurable variables that contextualize it among these three components. In the case of exposure to the climate, these might include the rate of temperature rise, the intensity of ultraviolet radiation, or sea surface current speeds. In the case of adaptive capacity, the variables might include the level of wealth, infrastructure, and education, while in the case of ecological conditions, the abundance of fish and corals may be most important.

We combine the most relevant variables for each component into a single axis. Thus, we develop an axis of key ecological conditions, an axis of exposure to climate change, and an axis of adaptive capacity, each determined by local conditions. Together, the three components help to contextualize key aspects of the social-ecological site. As we discuss in the subsequent chapters, appropriate adaptation efforts will depend on the interaction of these three axes, and a site's position along these axes can provide important information about the types of policies and actions required, which are discussed in the final chapters.

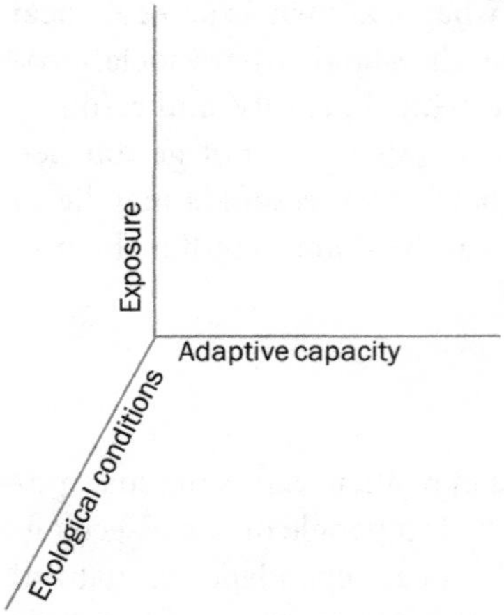

FIGURE 1.1 Conceptual model of management priorities and actions dependent on the location of social-ecological site within the three axes of environment, ecology, and society.

2

The Global Context of Marine Fisheries

The majority of this book is focused on the local and regional impacts of climate change to coastal ecosystems and societies. These local issues are, however, taking place in a dynamic global context of increasing population, increasing affluence in developing countries, and political inequity between the wealthier North (North America and Europe) and the poorer South (Africa, Southeast Asia, Latin America, the Pacific). Marine fisheries are an arena where the economic and political tensions around resource access and trade have become heightened. Fisheries resources have declined globally, and climate change impacts will change the distribution, productivity, and access (Worm et al. 2009). Consequently, this chapter highlights key global marine fisheries issues and trends to contextualize how climate change will influence the adaptive potential of our African-western Indian Ocean study region.

Global Trends and North-South Disparities in Fish Production

Increasing demand for seafood, expansion of fishing effort, and the globalization of both fish catch and markets makes the global context of fisheries increasingly relevant to the management of local-scale fisheries (Pauly 2008). For example, the total value of world exports of fish and fish products in 2006 reached U.S.$86.4 billion, which is a 55% increase from 2000 (Asche and Smith 2010). This global value is considerably larger than any other single potentially renewable agricultural export commodity. Coffee, for example, which is the second-most traded agricultural product, had a total export value of $12.3 billion in 2006 – just one seventh that of fisheries (Asche and Smith 2010).

The majority of this growth in the seafood trade is coming from fisheries in developing countries, and Africa is becoming increasingly central in this growth trend (FAO 2009). Fisheries play a huge role in employment, trade, and economics. It is estimated that 120 million people are dependent on fisheries for some part of their household incomes, with 35 million of these people in Africa (FAO 2009). However, the ecological consequences of the removal of fish at this scale are severe (Jackson et al. 2001, Watson and Pauly 2001).

WILD-CAUGHT PRODUCTION

The Food and Agriculture Organization (FAO) estimates suggest that global fisheries production has been relatively stable since the 1980s (FAO 2009). But a closer look reveals several trends that are a cause for concern. These include a profound shift in the source of fish as well as poor reporting in some countries, which can skew the global catch statistics.

First, grouping global catch data by wealthy (developed) and poor (developing) countries reveals a declining trend in catches from developed country from the 1980s (Fig. 2.1a). The total global production appears, however, to be maintained by a continued rise in catch in the developing nations. In the 1950s, wild-caught fisheries production from developed countries was about 15 million metric tons (MT) and about 75% of the global production. By the late 1980s, production from both developed and developing countries had each grown to 40 million MT. Yet, wild-caught fisheries production in developed countries declined by half after a peak in the late 1980s. Production from developed countries is now dwindling toward the volume of the 1950s. By 2007, the proportional contribution of developed and developing countries

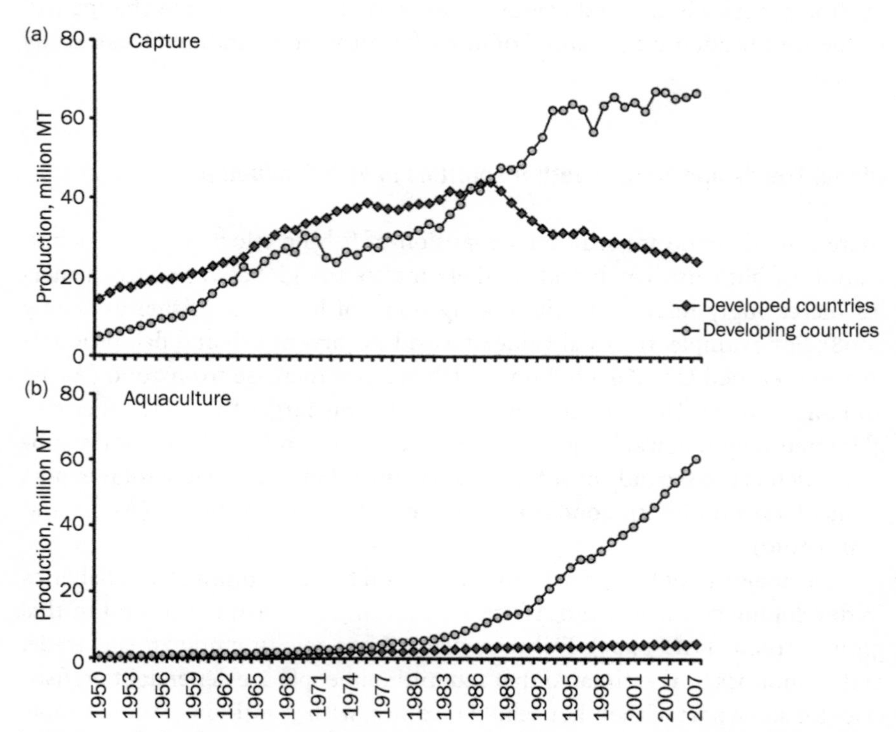

FIGURE 2.1 (a) Catch in million metric tons (MT) in developing and developed countries and (b) aquaculture trends.

Source: FAO (2009b).

had nearly reversed, with developing countries contributing around 70% of the total 110 million MT of global production. The net economic value that developing countries exported rose from $7.2 billion in 1986 to $24.6 billion in 2006 (FAO 2009). This decline in fisheries in the developed world is a result of harvesting beyond sustainable levels and the associated species and habitat losses, but in recent years this trend may also represent efforts to reduce fishing effort and rebuild some fisheries (McClanahan et al. 2008a, Worm et al. 2009). If properly implemented, this rebuilding could lead to a resurgence in developed world catches in the coming decades.

Second, poor reporting by some countries has masked substantial declines in global fisheries production. In particular, China consistently over-reports its fish landings data (Watson and Pauly 2001). When extreme over-reporting by China is corrected for, total global fisheries catch shows a steady decline from a peak of roughly 80 million MT in the late 1980s to about 70 million MT in 2000 (Watson and Pauly 2001). This is expected to substantially outweigh the widespread illegal and unreported catches (Pauly et al. 2005, Sumaila et al. 2006). Additional troubling trends in global wild-caught fisheries include a substantial decrease in the size of fish being caught in many parts of the world and an increasing depth at which fish are caught (Pauly et al. 2005). The latter point indicates that overfishing of local fish populations is likely being masked by expansion into new and deeper fishing grounds (Pauly et al. 2005).

EXPORTS

Interestingly, even though the developed world is producing less seafood, it is increasingly exporting a much greater percentage of its fisheries production (76% in 2007; FAO 2009). In contrast, the developing world exports only 28% of its catch (FAO 2009). There are also important qualitative differences in the types of fisheries products that are produced by developed and developing countries. Developed countries generally export lower quality products, such as nonedible fish meal and oils, that are used in developing countries, but they also export some of the very highest value products, including salmon, lobster, surimi, and caviar. Developed countries are able to maintain high values for exports by adding value in the production chain from the marine products landed to those exported. For example, the United States landed only roughly $4.4 billion in seafood, but its exports were worth $20.8 billion because of the four- to five fold increase in value in the production chain (NOAA 2009).

THE INCREASING ROLE OF AQUACULTURE

Aquaculture is compensating for some of the lost production of wild-caught fish. At 60 million MT/year, aquaculture makes a significant contribution to the total global seafood production (Fig. 2.1b; FAO 2009). Most of this aquaculture production comes from the developing world (FAO 2009). China produces nearly half of the global aquaculture production, but shrimp pond production in a number of tropical nations, such as Ecuador and Thailand, are also major contributors. Mainland China's aquaculture production is nearly twice their wild-caught production at 31 and 17 million MT, respectively, in 2007 (FAO 2009). Aquaculture production has the potential to supersede wild-capture fisheries as the primary source of fish production in a number of countries.

The combination of wild-caught seafood and aquaculture from the developing world appears to be compensating for the developed countries' increasing demand but declining production. Nevertheless, even though the total aquaculture production has increased, the percentage growth rate of finfish aquaculture production from key countries (Chile, Norway, Canada, UK) and globally has been declining since a peak in the mid-1980s (Liu and Sumaila 2008). These declining growth rates are a result of (1) scarcity of locations suitable for aquaculture; (2) declining market prices for key aquaculture species; (3) scarcity and associated high costs of inputs such as feed; (4) stricter environmental regulations, which increases the costs of compliance; and (5) increasing consumer awareness about the problems associated with aquaculture (Liu and Sumaila 2008). The declining growth rate of aquaculture makes it unlikely that aquaculture production will continue to compensate for the declining yields of capture fisheries.

It is also critical to consider the environmental impacts of aquaculture. Many forms of aquaculture require food derived from wild-caught fisheries. Many of these, such as salmon and tuna aquaculture, have very poor efficiency ratios, which means it can take up to 5 kilograms of protein from wild-caught fish to produce a single kilogram of protein from farmed fish (Naylor et al. 2000). In addition to increasing the pressure on wild-caught sources of fishmeal, aquaculture has also caused significant challenges to the sustainability of marine ecosystems through habitat conversion, disease, effluents, and escapees mixing with wild populations (Naylor et al. 2000). For example, shrimp exports from aquaculture in Ecuador were valued at over $1 billion in 2000. The following year, the export value dropped 80% due to disease outbreaks resulting from mangrove clearing and unsustainable farming practices (McClennen 2006). It is expected that the declining diversity of marine ecosystems, increased monocultures created by aquaculture, dependence of aquaculture on wild-caught fish, and climate change will converge to increase the vulnerability of the fisheries production systems (Worm et al. 2009).

Why Fisheries Are so Difficult to Manage

Wasted effort and resources caused by too many boats chasing too few fish has cost a staggering $2 trillion over the past three decades (World Bank 2006,

FAO 2008). The reasons for these losses are well known: specifically, perverse incentives and poor governance have created severe resource management and economic problems. These problems stem, in part, from some key characteristics of fisheries resources, which create incentives for overexploitation.

Fisheries are considered a common-pool resource. The defining characteristics of common-pool resources are that they have low excludability, meaning it is difficult to exclude people from accessing the resource, and high subtractability, meaning that when one person takes some fish, there is less for other people to take. Common-pool resources are particularly difficult to manage because these defining characteristics can create incentives for users to overexploit the resource in a situation known as a "tragedy of the commons" (Hardin 1968). Because it is difficult to prevent others from accessing "common" resources such as fisheries, users have incentives to catch all the fish before someone else does. Recognition of this problem has led to a number of efforts at many governance levels to rectify the situation, including a reexamination of the legal structure, which is partially responsible for the demise of marine fisheries.

International Efforts to Reverse Declining Fishery Trends

In response to the difficulty in managing common resources, legislation during the late 1970s and early 1980s reduced the extent of the global oceanic commons, which were replaced by an extended offshore jurisdictional zone controlled by coastal nations. To facilitate the harvest and management of marine resources, the United Nations Convention on the Law of the Sea (UNCLOS) extended coastal state jurisdiction from three nautical miles to 200 nautical miles (UNCLOS 1982, Articles 55, 56, 57). UNCLOS provides up to a 200-mile exclusive economic zone (EEZ) where coastal states have sovereign rights for the purpose of conserving and managing living resources and protecting the environment. The establishment of an EEZ delegated stewardship of areas to national governments when they were previously common property. UNCLOS empowers states to create and enforce protective measures within its territorial sea (Article 21) and creates obligations for states to protect the marine environment within their jurisdiction (Article 192, 194).

In 1995, FAO established a voluntary code of conduct that established a central paradigm of sustainability for fishing nations and suggested a list of appropriate and sensible guidelines. These guidelines are among the first statements sponsored by the United Nations (UN) on what is now commonly called ecosystem-based management. The code provides useful guidelines for establishing sustainable fisheries. The European Union's (EU) Common Fisheries Policy of 2002 also builds on UNCLOS by stating that legal recourse can be enacted on polluters (EU 2009). Some key elements of the EU Common Fisheries Policy are that (1) fisheries policies should be based on the principle that resources should not be used until there is evidence that the use will not be harmful, known as the "precautionary principle", (2) environmental damage should be rectified at the source, and (3) those causing environmental damage should pay and be held accountable through legal proceedings. The World Wildlife Fund (WWF), a conservation nongovernmental organization (NGO), has used this policy to encourage fisheries management, but legal actions against those causing environmental damages have not been successful.

The Johannesburg Plan of Implementation was based on the World Summit on Sustainable Development, adopted on September 4, 2002. The Johannesburg Plan contains a long list of recommendations for sustainability and encourages nations to adopt the FAO Code of Conduct. Both the FAO Code of Conduct and the Johannesburg Plan provide a good starting point to conceptualize and catalyze movements toward sustainability, but they may also be so comprehensive and demanding of financial and human resources that it is difficult for many nations, even wealthy ones, to implement them (Pitcher et al. 2009). The FAO Code of Conduct could be improved if costs and priorities were explicit so that nations could identify the most critical and affordable options.

These agreements acknowledged the need to manage fisheries on a global scale in a sustainable way, but none are legally binding. Consequently, although these arrangements provide guidelines and increase international pressure among nations to achieve similar sustainability goals, they lack clear accountability and mechanisms of enforcement and legal consequences beyond states' jurisdictional zones. They also lack clear and proven pathways to achieve sustainability among nations with highly variable financial resources for their management.

Implementing Fisheries Management at the National Level

Numerous other actions have also been proposed at the national level, most important, the early efforts by the United States in establishing the Magnuson-Stevenson Act in 1976, which was amended in 1996 and 2006. The amended versions include lessons learned in the evolution of the management process. This act established a legal framework for sustainable use of fisheries in the United States' EEZ. The act phased out foreign fishing in the EEZ that was established in 1983 and set up regional councils for the local management of fisheries; the final version promotes ecosystem-based management. Many national program and fisheries policies have similar legal and institutional frameworks often based on these examples from developed counties, but implementation remains a persistent problem that needs to be addressed in affordable and appropriate ways (Alder et al. 2010).

A multinational evaluation of the management of marine resources in 53 countries found that all were failing to fully implement strong management measures and that developing countries lagged behind developed countries (Alder et al. 2010). Top-performing countries such as Australia, New Zealand, the United States, and Germany were implementing measures to manage their marine resources sustainably. These efforts often include the establishment and financing of marine protected areas (MPAs), reducing or eliminating perverse subsidies, reducing trawling, minimizing fuel consumed in the fishing sector, and establishing sustainable seafood certification from the Marine Stewardship Council. However, the evidence for sustainability in the largest sense, including the ecosystem services, is often weak (Livingston and Sullivan 2007, McClanahan et al. 2008a). Sustainability of fisheries in large countries such as the United States can vary by state and local management. Current evaluations suggest that California's fisheries and some Alaskan species are not far from sustainable, while the states in New England have not recovered from a long history of excessive effort (Worm et al. 2009).

The costs of implementing the types of measures outlined in the FAO Code of Conduct might be one reason for these differences in performance between developed and developing countries (Alder et al. 2010). Some indicators of fishery performance, including marine mammal and seabird protection and status, use of fishmeal in aquaculture, and fisheries subsidies, are areas where developing countries were not disadvantaged. These countries may simply be unable to afford bad practices, such as subsidizing fisheries and expanding trawling. At the national level, there may be different pathways to managing and mismanaging marine resources dependent on the commitment and finances available for their implementation (Alder et al. 2010).

How overfishing in the EU is leading to an increasing dependence on imports of fish and fishery products is discussed in Box 2.1. Global trade policies and the fisheries contracts (discussed later in the chapter) encourage this continued need and reliance on trade and net importation. The North-South politics around these fisheries resources will further unfold in the coming decades. Along with climate change, these politics will greatly influence the wealth, adaptive capacity, and resilience of these respective regions.

Globalization and Distant-Water Fishing Contracts

Although many developed nations are attempting to rebuild their own fisheries, seafood production is increasingly occurring farther from central consumer markets, farther from state-controlled coasts, deeper into the sea, and

BOX 2.1

Increasing Dependence on Imports from the Developing World: A Case Study from Europe

European fish stocks are in poor condition, and efforts to rectify this situation are slow and lagging behind those of some developed economies, such as the United States, Australia, and New Zealand (Worm et al. 2009). Eighty-eight percent of Europe's 50 major fish stocks are overfished and 30% are classified as unsafe, meaning their potential to recover is jeopardized by their low abundance. In fact, since the initiation of the UNCLOS agreement in 1982, when management toward greater sustainability was initiated, stocks then evaluated as unsafe have continued to decline (Froese and Proelß 2010). This is largely because 91% of the species stock biomass is below the level at which fisheries yields are maximized (MSY), or fishing mortality rates are so high that their recovery to sustainable levels is not possible.

Beginning in 1999, a slight declining trend in fishing mortality rates occurred, such that if these declining mortality rates continue, 75% of the stocks would be fished at MSY by 2048 (Display 2.1). This rate of recovery is considerably delayed from the 2015 deadline established in the Johannesburg Plan. Further, a complete closure of the fisheries for 5 years would still not reach the targets by 2015 (Froese and Proelß 2010). This process of overfishing continues even though closure of the fishery could increase the 2009 yields of 7.6 to 13.6 million metric tons (MT) by allowing the biomass of stocks to increase from 31.8 to 51.6 million MT over a period of roughly 5 years. Increasing fish stock biomass would also reduce the costs of fishing and increase the stock's resilience to climate change. However, overfishing largely continues because of the capital invested in the fishery, unwillingness to exclude effort, the lack of political will, and the non-legally binding status of the various policies, codes, and plans.

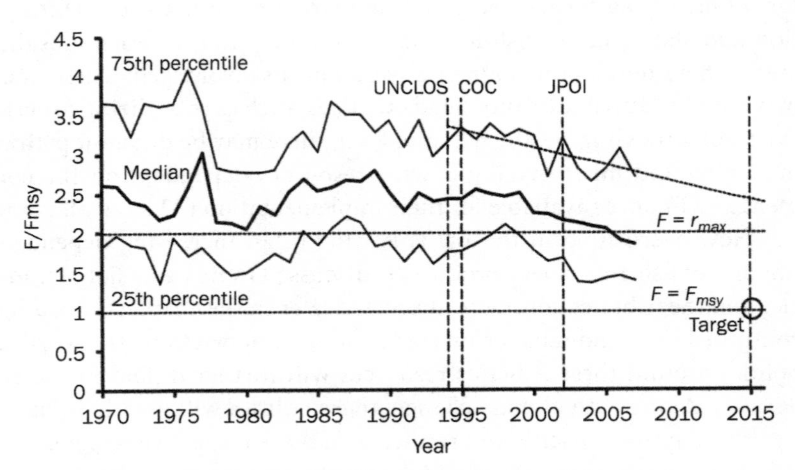

DISPLAY 2.1 Changes in mortality of the 50 common European Union fisheries stocks and the proposed target of maximum sustainable yield (MSY). The dates of key documents and agreements are shown, including the UNCLOS or Convention on the Law of the Sea, Food and Agriculture Organization (FAO) Code of Conduct (COC), the Johannesburg Plan of Implementation (JPOI), and the date when EU targets will be reached. R_{max} = intrinsic rate of population increase and F = fishing mortality.

Source: Froese and Proelß (2010).

Overfishing in the EU is leading to an increasing dependence on imported seafood. For example, in 2007, the EU imported $U.S.12 billion worth of fish and fishery products, accounting for 60% of its fish consumption. Europe exported $U.S.2.2 billion worth of fisheries goods in 2006, the bulk of it to large and high-end markets like Japan. In the Doha Round of World Trade Organization negotiations in 2001, the EU sought to reduce tariffs for European fisheries exports and tighten international rules on subsidies. The EU gives preferential access to their fish, fishery products, and markets through the Generalized System of Preferences where no tariffs are charged for fisheries imports from the 50 Least Developed Countries.

---

even into the EEZ of developing countries through contractual access rights and illegal fishing (Berkes et al. 2006, Pauly 2008). There are two main ways whereby this expansion is happening: (1) the globalization of markets and (2) the geographic expansion of developing country fishing effort in what is known as distant-water fishing (Swartz et al. 2010).

The globalization of fisheries markets occurs when local fisheries become linked to global demand for seafood (Berkes et al. 2006). The globalization of markets allows people to obtain resources from farther afield, often from areas that are poorer or less regulated. This process, however, is occurring throughout the world, in both developed and developing countries (Berkes et al. 2006, Worm et al. 2009). It often results in high market values being placed on marine resources that were formerly considered of little or no value. Global markets for marine products can emerge rapidly and deplete local marine resources before international or even national legislation is able to set up protection (Berkes et al. 2006). For example, Asian demand for live reef fish has resulted in serial depletion of specific reef fish species in parts of the Pacific before these countries are able to develop management responses (Scales et al. 2006, 2007). In some cases, incentives to exploit these newly commercialized resources are so great that they weaken locally managed institutions that govern marine resources (Cinner and Aswani 2007). For example, east African trade in sea cucumbers, ornamental snails, and aquarium fish were globalized some decades ago, but recently lower value octopus have begun being exported and can potentially be priced on the international market and become too expensive for local consumption.

Perhaps of more relevance to the geopolitical context of fisheries in Africa are distant-water fishing contracts. Distant-water fishing contracts or "bilateral fisheries partnership agreements" are the legal mechanisms that allow developed world countries to fish in the EEZ territories of developing countries, typically referred to as distant-water fishing (Fig. 2.2). The basic premise of these contracts is that the foreign fleets are utilizing surplus production of fisheries that host nations cannot access due to inadequate fishing capital. In principle, agreements such as those in the European Common Fisheries

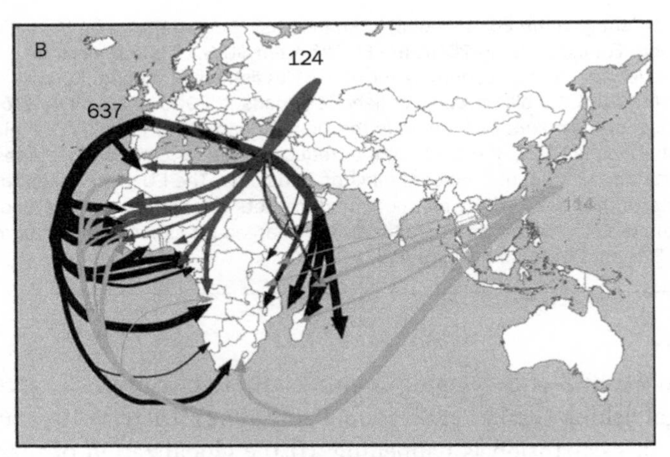

FIGURE 2.2 Number and distribution of foreign-fleet fishing contracts.
Source: Worm et al. (2009).

Policy are used to promote responsible and sustainable fisheries in the waters of non-EU countries. Contracts provide the European fleet with access to "surplus fish resources" in the territorial waters of African and Pacific countries and contract fees are expected to support the countries' fisheries policies. For example, in the mid-1990s, it was estimated that the cost of license fees was between 6% and 32% of the costs of fishing for European vessels fishing in West Africa (Porter 1997).

The contracts are attractive to recipient countries to finance national budgets, particularly fisheries programs, and also to provide jobs in the fishing and canning industries as described for the western Indian Ocean in Box 2.2. In West Africa, about one third of the total jobs created by these contracts are taken by African nationals with about 40% of these jobs in canning (IFREMER 1999). They also allow developing countries to earn hard currency, which can be used to pay off national debt (Watson and Pauly 2001). Developed countries benefit because these distant-water fisheries contracts help supply the fish deficit in developed countries, employ developed country workers, and serve as an outlet for the excess fisheries capital in developed countries that is resulting in suboptimal production. For example, Spain operated 750 vessels along the western African coastline during the 1990s.

Comprehensive evaluations of either the benefits or problems with distant-water contracts are generally lacking. Two key reasons for this are poor evaluation of fish stocks and catch reporting in host nations, and a lack of transparency over contract money by host countries. First, fish stocks in host nations are so seldom evaluated and illegal and unreported catches are so common that evaluations relating to improved management

BOX 2.2

Fishing Contracts in the Western Indian Ocean

The history of distant-water fishing in Africa started in the 1950s when the former Soviet Bloc caught approximately 75% of the total landings in the region, taking 1.5 million metric tons (MT) in the 1970s (Alder and Sumaila 2004). European fleets caught around 20% of the landing or just under 0.6 million MT until the 1990s and this later increased to 1.4 million MT or 55% of the landings. Asian fleets have caught the remainder of the fish taken (less than 11%). Distant-water fishing contracts, which allow foreign nations to fish another country's waters and which made these catches possible, are now expanding to the eastern coast of Africa and elsewhere. In the western Indian Ocean in 2008, the EU reported contracts with the Seychelles, Comoros, Madagascar, and Mozambique ranging from a low of U.S.$0.28 million in Comoros to a high of U.S.$3.8 million per year for the Seychelles.

The contracts state that 80% to 100% of the contract money allowing access the fishery should be used in support of the host nation's fisheries policies. These contracts are mostly part of a larger tuna network agreement that allows vessels mainly from Spain, Portugal, Italy, and France to fish in these waters. Madagascar, which received U.S.$0.86 million per year, also provides about 5% of the shrimp consumed by the EU. Mauritius ended EU contracts in 2007 and like some other countries it is trying to develop its own offshore fisheries capabilities. After gaining independence in 1990, Namibia chose to follow an incentive-based policy focused on quota taxation combined with tax reductions. This policy depended on Namibian national involvement in fisheries ventures and has resulted in a 1% gain in employment and a 1.5% gain in Namibian ownership for each percentage point loss in foreign fishing fees collected between 1993 and 1998 (Armstrong et al. 2004).

These distant-water contracts have evolved from simple access arrangements to agreements to make financial contributions that should support fisheries policy and include sustainable fishing as outlined in UNCLOS and the FAO Code of Conduct. In principle, money from these arrangements should support the creation of fisheries legislation for national and international fisheries, application of the laws, licensing, participation in regional fishery organizations, hiring of fisheries personnel, evaluating stocks and setting fishing effort limits, quotas, recovery periods, and other management responsibilities. The newer contracts are much more ambitious than those of the past but they still leave a number of issues unresolved, particularly regarding policy implementation, and this lack of enforcement undermines the proposed objectives (Kaczynski and Fluharty 2002).

Despite the economic and regulatory potential of distant-water contracts, they create problems: (1) they subsidize EU fleets, displacing local entrepreneurs in the host country; (2) they distort the economics of the European fishing enterprise by allowing a fleet that is disproportionably large relative to European resources; (3) they allow a substantial (up to 50%) loss of potential economic value in the host country; and (4) they promote excessive pressure on the resources, which undermines the long-term sustainability of host nation's fisheries (Sumaila and Vasconcellos 2000, Alder and Sumaila 2004). There is often lack of transparency on both sides of the agreement, such as EU countries not reporting excessive by-catch, underpayment of license fees, and failure to provide timely statistical information on catches (Kaczynski and Fluharty 2002). There are also some concerns that developed countries use these contracts as donor aid and to interfere in developing country national politics.

are difficult (Agnew et al. 2009). This lack of stock assessment seems particularly true for demersal fishes, which have the highest illegal and unreported catches, but may be less true for economically important species such as pelagic tunas (Agnew et al. 2009). Additionally, there are high levels of illegal and unreported catch in many host nations. For example, Agnew and colleagues (2009) found that the total estimated catches in West Africa were 40% higher and in the western Indian Ocean 10% to 40% higher than reported catches.

Agnew and colleagues (2009) found that the proportion of a country's reported catches to those that were illegal and unreported was strongly predicted by the regulatory quality, rule of law, control of corruption, and government effectiveness of the country, rather than the price of fish or the size of the country's EEZ. Similar findings were found for an evaluation of the adherence to the Code of Conduct, where most countries fail to fully comply and the frequency of noncompliance increases as the country's governance weakens (Pitcher et al. 2009). Consequently, more involvement in increasing incentives for good governance, such as preferential contracts and import arrangements for countries with good and improving governance, are needed by contracting nations to improve the chances that contract money is used toward their stated objectives of sustainability. Pitcher and colleagues (2009) suggest that improving compliance with the code will require establishing mandatory instruments, either national or international, and tailoring aid to address specific weaknesses.

Finally, it may be that more oversight of host nations is also needed to ensure transparency in the use of the contract fees toward achieving UNCLOS or FAO Code of Conduct recommendations. The use of this contract money is seldom clearly stated or monitored to ensure that the money is spent to increase sustainability of fisheries. Host nations could also improve contract conditions and retention of the economic value in their countries by coordinating negotiations for offshore fishing with their economies, to meet EU guidelines on fish handling and quality.

Donor Assistance for Fisheries Research and Management

Scientific understanding of tropical fisheries ecosystems, technological improvements to aquaculture, and improved capacity for governance and management are severely underfunded by donor aid relative to their importance to global economies as described earlier. The Organization for Economic Co-operation and Development (OECD), which is a group of 32 developed countries, reported that in 2008 a total of roughly $440 million was spent on official fisheries development assistance to non-OECD countries. Of this total, $180 million was spent on fisheries development; $158 million for administration, policy, and monitoring; $88.4 million for fishery services; and $12.3 million for fisheries research. The biggest donors—Japan ($151 million), the United States ($109 million), and Spain ($98 million)—account for 82% of this assistance to developing countries (OECD 2010). It is not surprising that these three OECD countries rely most on imported fish or fishing contracts. This expenditure, while small, is expected to ensure good relationships with countries contributing to the global fisheries production (Table 2.1). There are many nations that also rely heavily on fishing imports but contribute minimally toward foreign assistance.

When compared to the billions spent on fisheries imports and offshore fishing rights, these assistance expenditures are small, all less than 1.5% of the value of the OECD countries' fisheries imports (Table 2.1). The values are also small compared to the expenditures for management of their own national fisheries. For example, the U.S. National Marine Fisheries Service spent around $600 million managing domestic fisheries worth $4 billion, or six times the amount spent on U.S. foreign assistance in managing the $13 billion of imported fish. A reasonable premise of this discussion is that developed countries are currently attempting to improve the management of their own fisheries resources albeit slowly but, rather than eliminating effort, their excess fishing capital and labor is being used in poorer countries. At the same time, these developed countries are importing fish from countries with less investment in management, and while importing nations are providing minor assistance to exporting nations, it is considerably less than the costs of fishing and associated management in their own countries.

TABLE 2.1

Top ten nations in terms of fisheries catch, import value, and export value

Nation	Landings, million metric tons	Nation	Export value, billion dollars	Nation	Import value
China	17.3	China	9.0	Japan	14.0
Peru	9.4	Norway	5.5	USA	13.3
EU-27	5.6	Thailand	5.2	Spain	6.4
USA	4.8	USA	4.1	France	5.1
Chile	4.7	Denmark	4.0	Italy	4.7
Indonesia	4.4	Canada	3.7	China	4.1
Japan	4.2	Chile	3.6	Germany	3.7
India	3.5	Vietnam	3.4	UK	3.7
Russia	3.2	Spain	2.8	Denmark	2.8
Thailand	2.6	Netherlands	2.8	Korea Republic	2.7

Source: Based on 2006/7 figures from FAO (2008).

The North-South Disparity in Climate Change Impacts

Climate change is a compounding threat to the sustainability of capture-fisheries and aquaculture in many countries, but it may have a mix of positive and negative effects depending on geography (Cheung et al. 2010). As discussed in the following chapters, climate change impacts occur as a result of both gradual warming and associated physical changes as well as from the frequency and intensity of extreme events. Importantly, these changes take place in the context of other global socioeconomic pressures on natural resources and the geopolitical context of fisheries as described in this chapter.

Many marine species will shift their distributions toward the poles with climate warming, expanding the range of warmer water species and contracting that of colder water species. The most rapid changes in fish communities will occur among pelagic species and include vertical movements to counteract surface warming and maintain optimal body temperatures. Fish populations at the poles will extend their ranges and increase in abundance with warmer temperatures. Populations near the equator, on continental shelves, and in enclosed bays and seas will decline in abundance as temperatures warm (Barange and Perry 2009, Cheung et al. 2010).

These shifts in production will have implications for the production of fish in developing and developed countries. Cheung and colleagues (2010) projected the changes in global catch potential for 1,066 species of exploited marine fish and invertebrates from 2005 to 2055 under climate change scenarios (Fig. 2.3). A large-scale redistribution of global catch potential is predicted. An increase of 30% to 70% is projected in high-latitude or developed country regions and an overall drop of up to 40% is projected in the tropics, largely in developing country regions. Exclusive economic zone regions with the

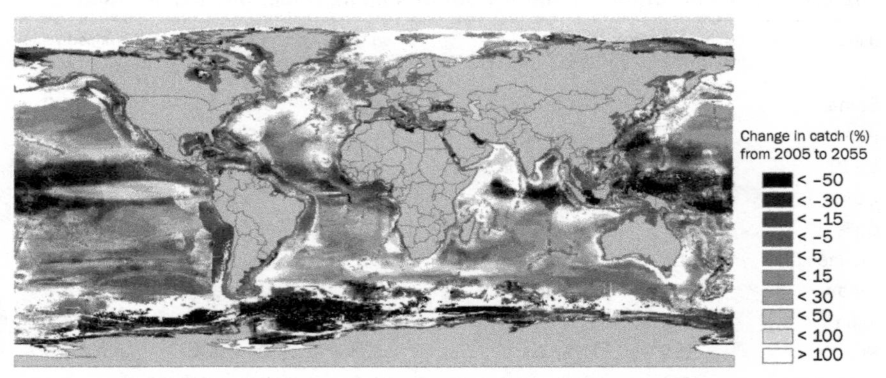

FIGURE 2.3 Climate change effects on projected fish global catches for Intergovernmental Panel on Climate Change (IPCC) emission scenario of 720 ppm of CO_2 by 2100.

Source: Cheung and colleagues (2010).

highest projected increase in catch potential included Norway, Greenland, Alaska, and Russia; EEZ regions with the biggest potential losses included Indonesia, the continental United States, Chile, and China.

There is, however, significant regional variability in these predictions, and the tropics have some innate resilience in terms of having more species and smaller body sizes, which can provide more and faster options and responses to climate change (Fisher et al. 2010). Furthermore, predictions for the western Indian Ocean region are for increases in open pelagic areas but declines along the east African coastlines from Kenya to Mozambique and western Madagascar. Coastlines with potential increases in catch are eastern and northern Madagascar, the Seychelles, and Somalia. The overall Indian Ocean projections are different from most of the tropics with a 20% potential increase from 2005 levels under the high–carbon dioxide emissions scenario. Nevertheless, declines in catch potential are largely predicted for the northward boundary of the Indian Ocean and east African coastline.

Many of the affected tropical coastline regions are also socioeconomically vulnerable to these changes and will be challenged by both the prospect of lost fisheries productivity near-shore and the potential to compensate for this loss by increased offshore fisheries productivity. It is often these offshore fisheries that are the targets of the distant-water fishing contracts, so ultimately, it may be very difficult for small-scale fishers and coastal societies in the western Indian Ocean region to directly benefit from this projected increase in pelagic production.

Projected reductions in catch potential in tropical countries and the increase in high-latitude countries coincide with similar projections for land-based food production systems (Easterling et al. 2007). The costs in capital, operating expenditures, and carbon emissions to access this offshore fishery are expected to increase and, combined with the shifts in catch potential, they will create challenges for meeting the trade and protein requirements of developing countries. The unfolding recognition of unequal distribution of resources and costs to access fish, management effectiveness, and the impacts of climate change are expected to lead to continuous changes in aid and the interactions between national and international governance of marine fisheries. The alignment of national and international interests and policies will influence the evolving framework for these institutional changes. How local laws are implemented, their flexibility, the effectiveness of monitoring resources, and mechanisms for funding will increasingly solve or undermine these emerging governance problems (Young 2010). The remainder of this book focuses on the African-Western Indian Ocean region and evaluates climate change and potential effects on the near-shore fisheries ecosystems and the adaptive options for the people of this region.

3

Climate Change and Oceanography

Climate science has produced an extremely detailed record of climate change over the past 1 million years (Alley 2000). In many cases, resolution of these studies is accurate to a few years, revealing abrupt changes taking less than a decade, which can have climatic consequences that can last for millennia (Steffensen et al. 2008). The last 2.5 million years has been characterized by glacial cycles that are caused by small and slowly changing variations in the tilt of the Earth on its axis and its rotation around the sun. The gross climate variations are cyclical and produce oscillations at frequencies of about 26,000, 41,000, 100,000 and 400,000 years. Empirical research on climate patterns over the past few million years is based largely on ice cores from Greenland and high-mountain glaciers. Thus, the record of climate changes in the tropics is not as long, resolved, and clear as in the temperate and arctic regions. Nonetheless, a coherent picture of tropical climate variability is beginning to emerge from other temperature proxies. This chapter briefly describes the current understanding of the recent geological history of climate and oceanography in the Indian Ocean and its implications for human-induced climate change.

The Last Few Millennia

Our understanding of the climate over the past 12,000 years is becoming clearer from a mix of tropical glacial ice cores, cave stalagmites, coral skeletons, and lake and ocean sediments. The planet entered a warm period more than 11,000 years ago and glacial conditions declined, as measurements from the arctic show two abrupt stages of rising temperatures, one of about 10.4°C and another of 4.5°C; these may have occurred over a period of just a few years (Kobashi et al. 2008, Steffensen et al. 2008). Since 11,270 years ago there has been a more stable climate but with some abrupt changes. For example, in recent times, four large-scale droughts ranging from 2 to 10 years have occurred since the 17th century in the Indian Ocean region (Cook et al. 2010). Temperature changes in the tropics are usually a little more than one half of those changes recorded at the poles but have strong effects on monsoon circulation strength and associated rainfall (Correge et al. 2004).

Between 11,000 and 5,000 years ago, east Africa was warmer and wetter than at present, despite some short-term but significant temperature changes. Oxygen isotopes extracted from stalagmites in caves in Oman indicate that drying was slowly under way in the northern Indian Ocean region from 8,000 to 4,000 years before the present (BP; Fig. 3.1). This drying was associated with a gradual southward migration of the inter-tropical convergence zone (ITCZ) and gradual weakening of the monsoons in response to Earth's orbitally induced declining northern summer sunlight (Fleitmann et al. 2003).

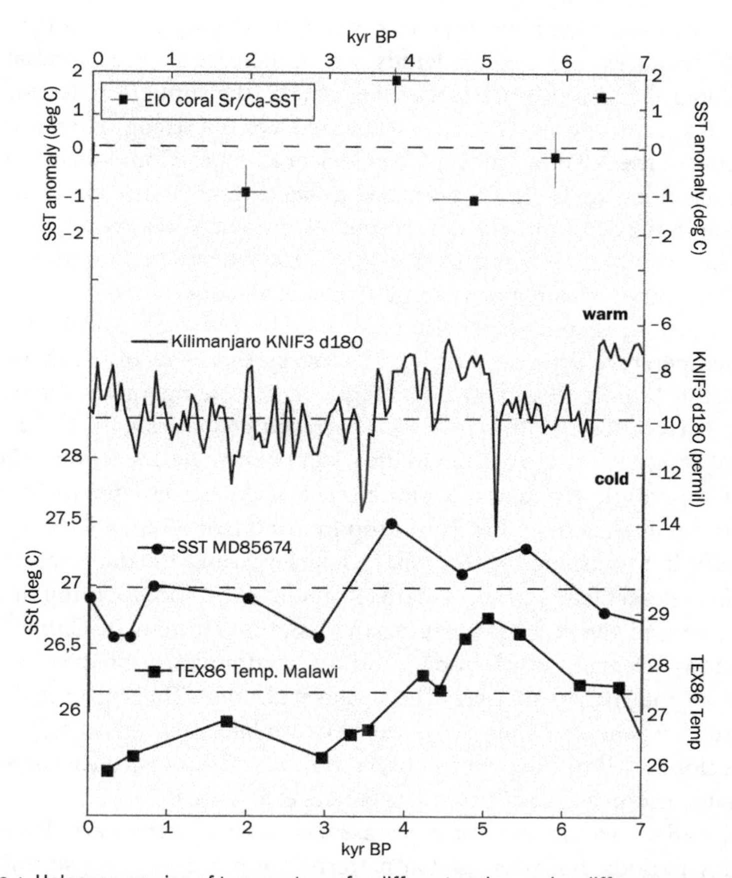

FIGURE 3.1 Holocene proxies of temperatures for different regions using different sources of data. The top is a comparison of sea surface temperature (SST) anomalies inferred from coral Sr/Ca (filled squares) through the Holocene in the eastern Indian Ocean (EIO) corals off Sumatra (Abram et al. 2008). The middle is an ice core $\delta^{18}O$ record from mount Kilimanjaro used as a proxy for east Africa air temperature (Thompson et al. 2002). Sr/Ca and $\delta^{18}O$ are chemical proxies for water temperature. Below that is an alkenone derived SST reconstruction from a marine sediment core from the western equatorial Indian Ocean (core MD85674 at 3°N, 50°E; Bard et al. 1997). The bottom is a lake sediment record from Lake Malawi (Castañeda et al. 2007) as a proxy for east Africa lake temperature (TEX86).

Stalagmite isotopes from southern China suggest that the last 9,000 years in Asia were punctuated by eight weak monsoon events that lasted 100 to 500 years and were not entirely explained by changes in sunlight. Changes were more likely associated with ice-rafting events that cool the North Atlantic. These in turn change the strength of mid-latitude westerly winds reaching Asia, shifting the monsoon position south, ultimately leading to less monsoon rains reaching China (Wang et al. 2005). Similarly, a dry period in Kilimanjaro glaciers and China's caves around 8,200 years ago was associated with weakened monsoons and extreme drying in Asia and eastern Africa (Thompson et al. 2002, Wang et al. 2005). One hypothesis to explain this drying pattern is that a cold Atlantic produces cold westerly winds, which increase snow accumulation in the Tibetan Plateau; this increases solar albedo (the sunlight reflection from the earth's surface) from the snow, which reduces the strength and northerly migration of the Asian monsoon (Overpeck et al. 1996). Consequently, Indian Ocean monsoon variability is predicted to depend on changes in snow accumulation in Asia and the Tibetan Plateau. Additionally, the recent increase in air pollutants or aerosols is also reducing surface temperature and monsoon rainfall in some Asian regions (Ramanathan et al. 2005).

Another cool and dry period starting in 6500 BP culminated in a rapid temperature drop and drying at 4200 BP that is recorded throughout Africa, the Middle East, and Asia (Fig. 3.1). This temperature change and drying is recorded in Kilimanjaro glaciers and coincides with the development of irrigation and the hierarchical agrarian societies around the Nile and Mesopotamia rivers and abandonment of desert settlements in Arabia (Thompson et al. 2002). This third drop in rainfall at around 4000 BP lasted 300 years; it is associated with drops in lake levels and the dominance of the associated desert fauna in the Sahara (Kröpelin et al. 2008), drying in southern China and the end of pre-agrarian Neolithic cultures of China (Wang et al. 2005, Dearing et al. 2008), and an accumulation of dust in South American and African glaciers (Thompson et al. 2000, Thompson et al. 2002). The late Holocene temperatures in east African lakes and the western Indian Ocean closely follow the pattern observed in the Kilimanjaro ice cores (Bard et al. 1997, Thompson et al. 2002, Castañeda et al. 2007).

Globally, warm temperatures are associated with wetter periods and cold with dry periods, but these global patterns are not always consistent in all regions (Mann et al. 2009). For example, lake fossils and lake-level reconstructions of Lake Naivasha, Kenya, suggest that periods of low solar input are associated with wetter conditions, and higher intensity of sunlight is associated with drought periods over the past 2,000 years. These wet and dry times have also been recorded in the oral histories of the people from the region (Verschuren et al. 2000). Warm periods in temperate Europe, known as the Medieval Warm Period, were associated with droughts in east Africa. Conversely, the Little Ice Age of Europe, from AD 1400 to AD 1850, was associated with largely wetter conditions in east Africa that declined after about 1880 (Verschuren et al. 2000, Lamb et al. 2003, Lamb et al. 2007). These observations indicate that rainfall in east Africa can be the opposite of larger global or ocean-basin climatic changes and one reason why coarse scale global climate models may not have strong predictive power in portions of the east Africa region (Brown and Funk 2008, Mann et al. 2009).

Ocean proxies indicate that upwelling strength declined off Somalia from 10,000 to 400 years BP, but after that this trend reversed and now closely follows the warming of the southern Tibetan plateau (Anderson et al. 2002, Thompson et al. 2003). Based on coral cores located throughout the Indian Ocean and a time series starting in AD 1660, Gong and Luterbacher (2008) determined that the cross-equatorial winds and monsoons were strongest in the late 17th to late 19th century and have weakened in the 20th century. Additionally, decreasing dust concentrations in Tibetan Plateau snow in the past 50 years indicate reduced winds over Tibet (Xu et al. 2007). The loss of extensive forests 1,400 years ago in China (Dearing et al. 2008), 400 years ago in India (Caner et al. 2007), and in the past 70 years in east Africa (Fleitmann et al. 2007) have probably contributed to dryer monsoon conditions (Knopf et al. 2008). Forest vegetation increases evapotranspiration and rainfall (Zhang et al. 2001), and smoke from forest fires decrease local rainfall (Rosenfeld 1999, Rosenfeld et al. 2008). These studies suggest that the monsoon rains are weakening and are likely to create more droughts in portions of east Africa and southern Asia, despite the warming ocean.

The Last Few Centuries

THE INSTRUMENTAL RECORD

The Hadley Center in the UK and the National Oceanographic and Atmospheric Administration (NOAA) in the United States have maintained empirical records and reconstructed estimates of ocean temperature in the Indian Ocean for about 120 years at monthly intervals with 111 × 111 km resolution. The records for the past 50 years are the most reliable. Since 1981, Earth-observing satellites and permanent monitoring buoys have further increased the frequency and spatial resolution of ocean temperature measurements to more than daily and to a 20 × 20 km resolution for the early records. Since 2000, the spatial resolution has increased to 4 × 4 km. This higher resolution is good for many purposes, but the longer time scales are critical for separating oceanographic oscillations from longer trends, such as human-induced global warming, where the signal is most observable from the mid-1970s (Ihara et al. 2008).

Surface ocean temperature data for the past 120 years in the Indian Ocean show the rising temperatures that characterize many of the world's oceans and specifically the increasing strength of warm oceanographic oscillations from the 1980s (Fig. 3.2). Observations suggest that decadal Indian Ocean temperatures vary in step with global temperature (Funk et al. 2008). A seawater temperature rise of roughly 1°C has been reported from a summary of the instrumental data since 1880. This rise has been variable in the western and eastern Indian Ocean, and over time (Ihara et al. 2008). During the period 1880–1919, the warming in the downwelling areas off east Africa was faster than the eastern upwelling area off Sumatra. Both sides began warming after that period, followed by a brief cooling, and after the 1950s there was a strong warming of 0.5°C in 50 years. The reported cooling during the 1940s is

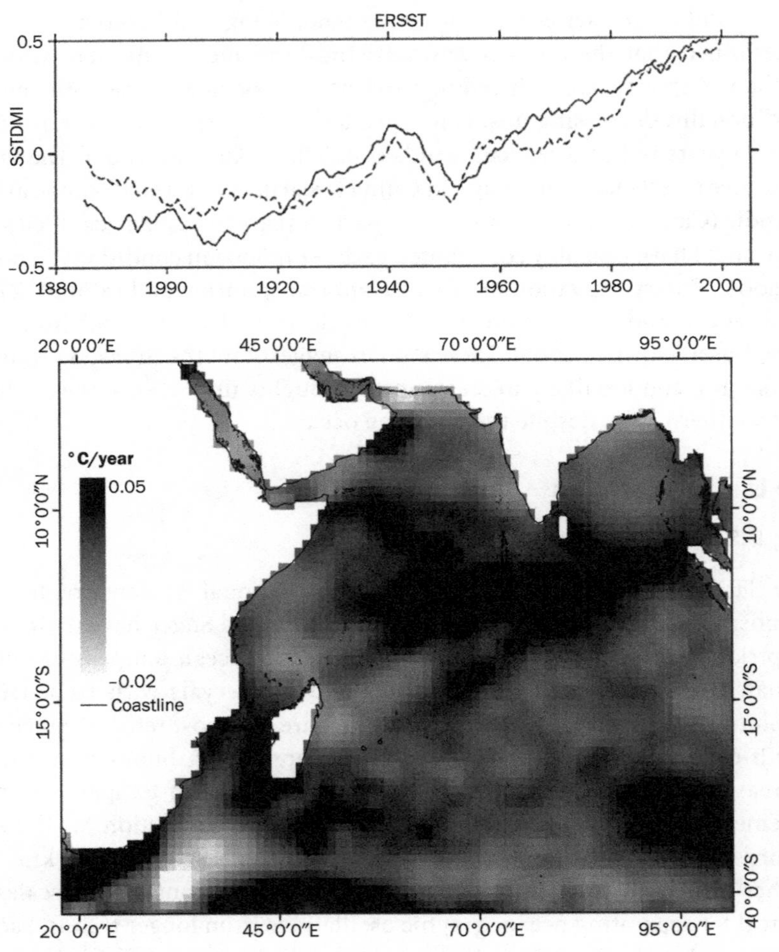

FIGURE 3.2 Surface seawater temperature (SST) changes (top) since 1880 based on 10-year running means for the eastern (dashed lines) and western (solid lines) Indian Ocean.

Sources: (top) Ihara et al. (2008) based on National Oceanic and Atmospheric Administration/National Climatic Data Center (NOAA/NCDC) Extended Reconstructed SST (ERSST) dataset, version 2; and (bottom) the 111-km grid rate of SST rise based on the monthly data for the Hadley 1950–2008 time series. *Sources:* Hadley Center (2009).

likely to be an artifact of the changing measurement instruments rather than a real change in temperature (Thompson et al. 2008).

TEMPERATURE PROXY ARCHIVES

High-resolution coral records can have monthly resolution and date back several centuries. These temperature proxies have the potential to provide a tool for detailed reconstruction of past surface ocean variability under periods of recent climate change. Proxies for seawater temperature from the skeletons of living corals are now available using stable isotopes ($\delta^{18}O$, $\delta^{13}C$) and trace metals (Sr/Ca) in calcium carbonate cores for a modest number of sites in the Indian Ocean. These coral proxy records estimate temperatures for the past 300 years, and the records all show rising seawater temperature during the past 100 years; they largely confirm the instrumental record in terms of the rate of rise and the importance of inter-annual oceanographic oscillations, such as the Indian Ocean dipole (IOD) and El Niño southern oscillation (ENSO; Fig. 3.3). Interestingly, coral proxies do not record the decline picked up by instruments in the 1940s, supporting the contention that changes in the instruments used to measure seawater temperatures were the cause of the reported change mentioned earlier.

The two dominant inter-annual oscillations in the Indian Ocean are the globally important and widespread ENSO and the regionally important IOD. Both oscillations have similar frequencies of between 2 and 8 years. While they are often independent, at times they coincide, as in 1998 (Fig. 3.3). This has made it difficult to separate the two fluctuations from proxy records of temperature, but the use of upwelling and river discharge proxies off Sumatra and northern Kenya, which detect climate signals during the active IOD seasonal window, have made it possible to separate the signals (Kayanne et al. 2006, Abram et al. 2008). The strength of the IOD has increased since the 1920s, replacing the ENSO as the dominant oceanographic oscillation in the tropical Indian Ocean (Nakamura et al. 2009). In addition, the frequency of the IOD has increased from about 10 years in the early part of the 20th century to 1.5 to 3.0 years since the 1960s and is associated with an increased coupling with the Indian monsoon and October to December wet seasons in east Africa (Bowden and Semazzi 2007, Abram et al. 2008, Nakamura et al. 2009). The increasingly stronger and more frequent movements of warm water from the eastern to western equatorial regions of the Indian Ocean can stress and kill temperature-sensitive marine species.

As the number of coral cores taken in different countries in the region increases, there are improvements in understanding the spatial variability in the climate patterns and distinguishing what are region-wide versus local responses. For example, the oldest proxy records from Mafia, Tanzania, extend the seawater temperature record back to 1622 (Damassa et al. 2006),

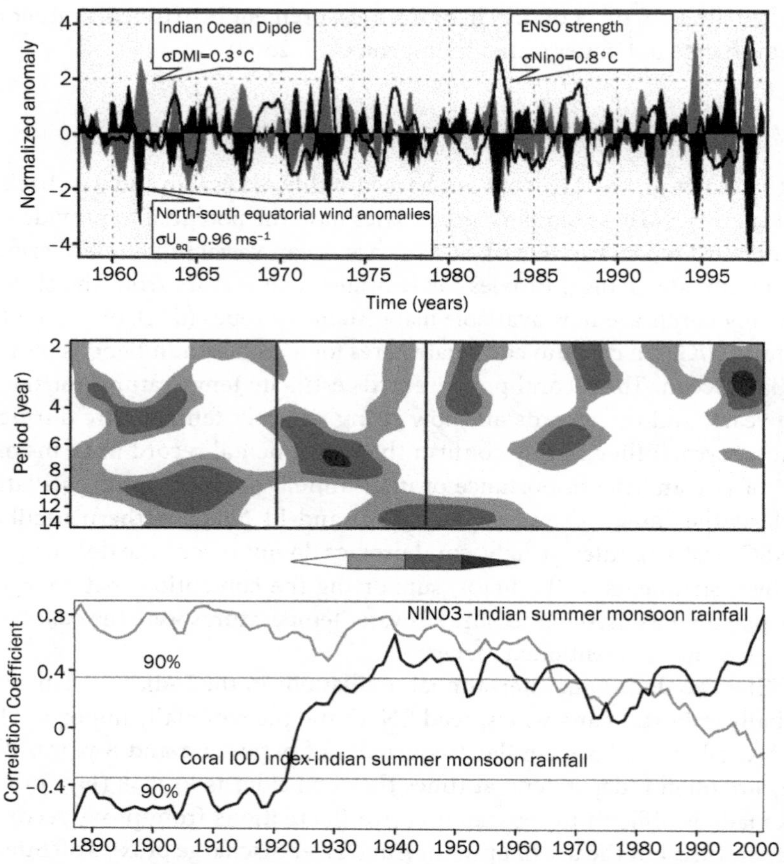

FIGURE 3.3 Reconstruction of Indian Ocean oceanographic oscillations. (Top) the relative strengths of the El Niño Southern Oscillation (ENSO) and Indian Ocean dipole (IOD) from 1955 to 1999. (Middle) periodicity of the IOD where darker colors indicate stronger periodicity, and (bottom) strength of the ENSO and IOD since 1885 based on correlations with Indian summer rainfall.
Sources: Saji et al. (1999) and Nakamura et al. (2009).

and another record from southwest Madagascar extends to 1658 (Zinke et al. 2004). Both of these old cores and a shorter core from the Red Sea indicate declining coral growth rates in the most recent 50 years (Cantin et al. 2010). The Mafia coral did not correlate well with recent climate variables; the authors suggest that this resulted from environmental stress after the 1950s, which may be associated with locally changing climate conditions including increased sediments and nutrients, high temperatures, or changing ocean chemistry. Prior to 1950, the Mafia coral tracked climate variability well and temperature oscillations of 2.5, 5–6, and 10–11 years were evident from 1896. The dominant oscillations found in the southwest Madagascar coral were on the frequency of 4 years and the inter-decadal band of 16–20 years (Fig. 3.4). The inter-

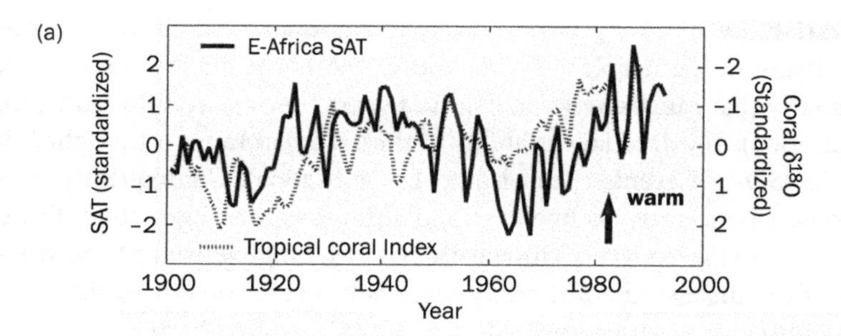

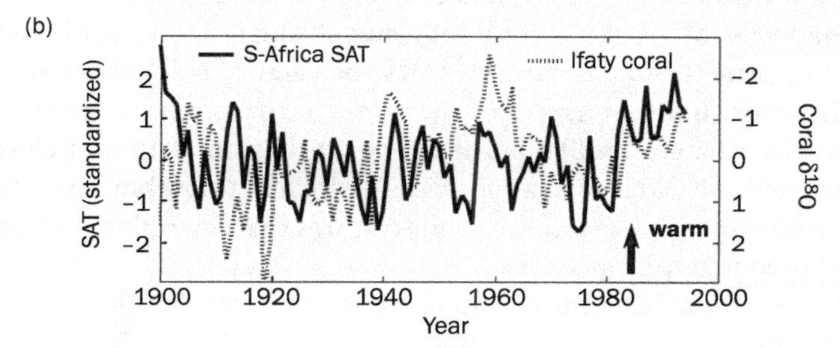

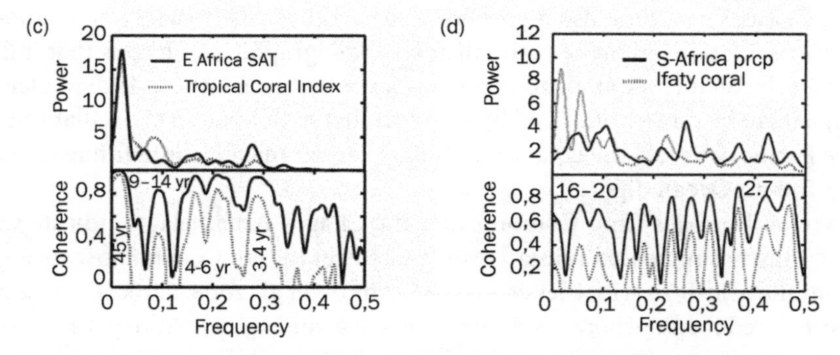

FIGURE 3.4 (a) Time series of annual mean surface air temperatures (SAT) over east Africa (10°N–15°S, 20–45°E; HadISST) compared to annual mean tropical coral $\delta^{18}O$ Index (Mayotte, Seychelles, and Malindi corals) averaged between July to June. (b) Time series of annual mean (March to February) SAT over southeast Africa (15–25°S, 20–45°E; HadISST –Hadley Center sea surface temperature estimates) compared to mean annual Ifaty (southwest Madagascar) coral $\delta^{18}O$. Negative anomalies in coral $\delta^{18}O$ isotopes correspond to warm SST and arrows indicate these warm anomalies. (c) Cross-spectral analysis of east Africa SAT with tropical coral $\delta^{18}O$ Index and (d) cross-spectral analysis of southern Africa rainfall with southwest Madagascar (Ifaty-4- coral core taken in Ifaty Bay) coral $\delta^{18}O$ for December to March season to determine the significant frequencies in the climate oscillations. Main frequencies significant above the 90% level are indicated by the year or range of years of the significant frequency. Time series were normalized by subtracting values from the mean and dividing by their standard deviation.

Sources: Zinke et al. (2009).

decadal variability of 16–20 years was also found in land rainfall records across southeastern subtropical Africa and is distinct from the tropical decadal variability that ranges between 9 and 14 years (Reason and Rouault 2002, Zinke et al. 2009). The decadal variability in the 9–14 year range is thought to be related to ENSO-like events (Reason and Rouault 2002). Consequently, there appears to be a large-scale decline in the conditions for coral growth but more local variability in the impact of climate oscillations, ranging from 2 to 20 years.

Both of the discussed coral proxy records indicate cooler conditions for the 17th-century time series; for instance, a 0.6°C cooling between 1622 and 1722. The southwest Madagascar coral indicates strong oceanographic oscillations of 3 to 5 years that did not closely track the Pacific ENSO between 1675 and 1730. The most unusual aspect of the 17th-century temperature record from Mafia was a longer oscillation of 20–25 years, similar to and closely tracking the Pacific decadal oscillation (PDO), which is a slow but important oscillation in North Pacific sea surface temperatures that can influence global weather and oceanographic patterns.

A coral core from Mayotte also indicates 5- to 6-year frequencies and decadal variability of 18 to 25 years associated with the PDO. The strongest PDO (18–25 years) signal in the last 120 years was found for the southwest Madagascar core and also a core from Reunion corals (Crueger et al. 2008); these oscillations influence rainfall and oceanographic conditions that influence fish catch (Jury et al. 2010a). Consequently, the southern Indian Ocean appears to be more influenced by or associated with long-term oscillations of the Pacific, whereas the equatorial Indian Ocean may be more influenced by the Indian Ocean dipole.

Coral proxies from Comoros and the Chagos Archipelago indicate that warming has increased rainfall over the central Indian Ocean, most strongly from the mid-1970s (Pfeiffer et al. 2006, Abram et al. 2008, Zinke et al. 2008) but has reduced onshore moisture transports and increased dry-air subsidence across eastern portions of the Greater Horn and areas from Comoros south, exacerbating the strengths and impacts of ENSO and IOD oscillations, especially during the March–May and June–August rainy seasons in Kenya and Ethiopia, respectively (Funk et al. 2008, unpublished data).

Looking Forward

MODELS

Models are frequently used to examine the likely impacts of climate change. On the large scale, climate models predict increased drought for northern and southern Africa and reduced drought for eastern Africa with climate warming (Arnell 2006). Empirical rainfall data from the last 20 years would appear to support the predictions for southern Africa, but rainfall gauge data reflect declining rainfall during the March–May rains since the 1980s (Funk et al. 2008). Similarly, the Indian subcontinent is experiencing declines in rainfall in about two thirds of the studied regions associated with increasing human-produced black carbon aerosols (Ramanathan et al. 2005). The warming Indian Ocean appears to result in more rainfall on the Indian Ocean but stronger high altitude westerly winds that keep the rain from reaching eastern Africa and black carbon aerosols that reduce the land-sea temperature gradient with India. Consequently, current evidence suggests that more rain will be experienced by central Indian Ocean island nations but not the eastern or southern African continental margin.

A warmer world is projected to produce more intense tropical storms in the southern Indian Ocean (Macdonald et al. 2005), and while the models did not predict these, more large, violent storms are emerging in the northern Indian Ocean, associated with weakened mid-latitude and high-altitude easterlies (Brahmananda Rao et al. 2008). A 2°C–4°C sea surface temperature rise is expected to increase cyclone intensity between 10% and 20% (Lal 2001), but this prediction depends on other factors, such as changes in wind shear at altitude.

The ocean has also begun to experience changes in productivity, with increases in the northern and decreases in the southern Indian Ocean (Gregg et al. 2003). Productivity increases that are often in the upwelling zones of northern Somalia and near-shore areas around Arabia and India may increase pelagic fisheries, but among coral islands of the central Indian Ocean, increases in planktonic productivity can reduce reef benthic productivity and diversity by reducing light penetration to the bottom.

Despite the occasional disparity between climate models and empirical data, the sum of this information indicates that the predicted anthropogenic climate change processes are under way and that climate models are most likely to be conservative or to underestimate potentially detrimental impacts (Funk et al. 2008, Richardson and Poloczanska 2008). Models often have difficulty predicting localized rainfall because of weakness in interpreting the hydrological cycle; difficulty in accounting for complex landforms, interaction between the Indian Ocean and the continent, dust and aerosol concentrations, and plant-climate interactions; and limitations in simulating high-altitude teleconnections and feedback mechanisms (Allan and Soden 2008, Funk et al. 2008). Climate models are not meant, at this point, to predict fine-scale resolution such as the 26 climatic zones in east Africa (Ogallo 1989). Nevertheless, the sophistication of data and models is constantly increasing and uncovering climate dynamics, and can inform the most appropriate adaptation measures.

Spatial Patterns

A critical point from both the historical records and the models is that the impacts of climate change will not be spatially uniform. Despite the heterogeneity, oceanographic and atmospheric processes have deterministic elements that will create some predictability to the patterns. For example, we have mentioned the east-west gradient in the tropical Indian Ocean formed by gross easterly movements of winds and currents. There are also differences in the tropical and southern Indian Ocean in terms of the frequency of oscillations as well as island and continental-current effects, such as the upwelling and downwelling. The remainder of this chapter highlights the key spatial patterns for both terrestrial and oceanic stresses. Chapters 8 and 10 further detail how knowing where a site lies in terms of oceanic and climate exposure can be used to inform policies, plans, and actions.

TERRESTRIAL SPATIAL PATTERNS

Rainfall is critical to plant growth and is one of the main limitations on agricultural production in tropical eastern Africa, although crop selection, fertilizer use, seed varieties, and the area that is farmed can influence this production. Consequently, changes in rainfall and how farmers respond are critical to food production and security (Brown and Funk 2008). In east Africa, there are two critical rainfall seasons: the long rains (March to May), which provide more rainfall and create conditions for the majority of agricultural cropping; and the short rains (October to December), which is a crucial, but less intensive rain period. The number of undernourished people in eastern and southern Africa has increased in the past three decades along with declines in the duration of the long rains and increasing population growth (Funk et al. 2008).

Food production models are predicting declining food production in much of Africa, but they have not fully accounted for observed rainfall declines that are greater than those predicted by climate models (Fig. 3.5). The projected outlook is suggesting further declining long-rain rainfall trends that may be compensated for in a few areas by increased rainfall during the short rains (Brown and Funk 2008). For example, continued drying is expected in most of Ethiopia, especially in the southwest during the March–September rains (Verdin et al. 2005). In contrast, northern regions of east Africa may experience modest rainfall increases in the short rains but not the southern regions, because the warm IOD has greatest influence on the equatorial regions.

Beyond changes in mean rainfall, the number of extreme climatic events is likely to be one of the main effects of climate change, with an increase in the extremes of wet, dry, and hot seasons. Based on three global climate models with two likely carbon emission scenarios, the number of extreme events expected at the end of the 21st century has been mapped (Baettig et al. 2007; Fig. 3.6). All Indian Ocean regions are expected to have more extreme events, with the lowest changes of 0 to 6.5 events above the current baseline in most of India and western Australia, but more extremes in tropical regions ranging

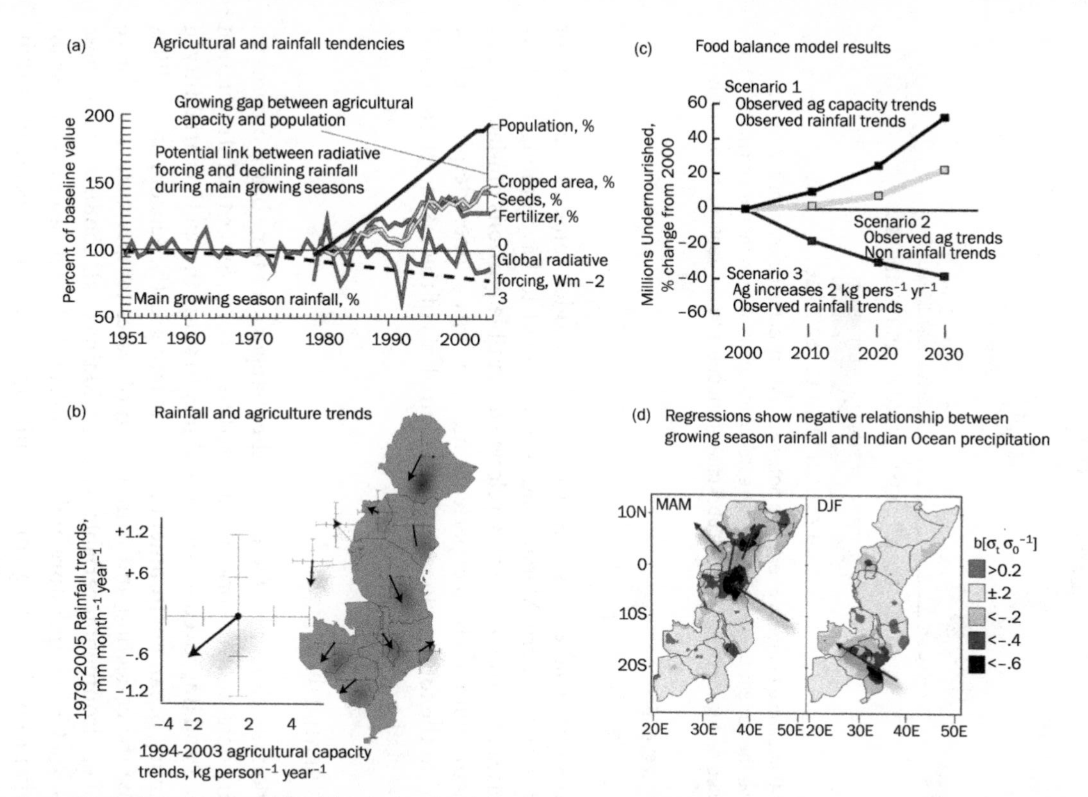

FIGURE 3.5 Food security analysis showing (a) the historical and projected trends in growing season rainfall, (b) the current vectors of rainfall and agricultural capacity by country, (c) a food balance model, and (d) association between rainfall in the two rainy seasons with Indian Ocean precipitation.

Source: Funk et al. (2008).

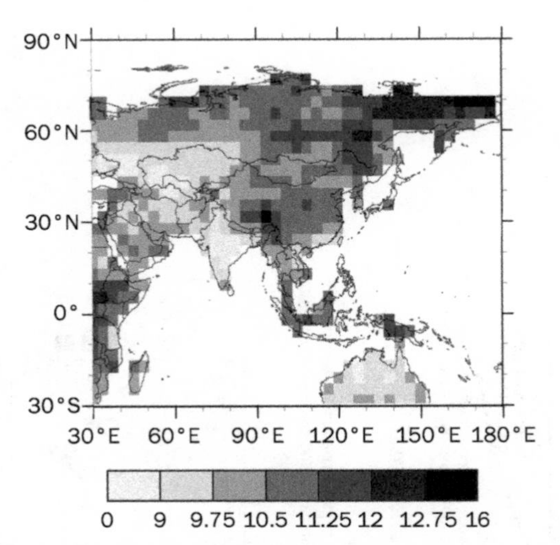

FIGURE 3.6 Cumulative numbers of extreme hot, wet, and dry annual events for a 20-year period above the current baseline. These extreme events are projected for the end of the 21st century based on the average of three global change models and two realistic carbon emission scenarios.

Source: Baettig et al. (2007).

from 7 to 8.5 unusual events. According to this model, some coastal areas of east Africa and eastern Madagascar are predicted to have among the fewest increases in extreme events. Cyclone-susceptible portions of Madagascar, however, are more likely to experience more severe cyclones, despite the lack of a trend in the total cyclone days for an analysis of the 1981 to 2006 period (Kuleshov et al. 2008).

OCEANIC SPATIAL PATTERNS

The previous sections have focused on the changes over time based on proxies from point data, such as coral, ice, and sediment cores. Since the introduction of satellites and improved sensors, the ability to study the spatial variability of this changing ocean environment has increased greatly. Sensors on satellites now allow the collection and compilation of water temperature and a large variety of environmental variables (Fig. 3.7, Maina et al. 2008). Many of these variables, such as temperature and ultraviolet (UV) light can cause stress for many organisms, including corals. Similarly, water flow speeds, photosynthetically active radiation (PAR), and nutrients interact with these stresses and are also important for biological production processes. The distribution and interaction of these environmental variables will influence the stress that species experience with climate change and their potential to adapt to the changing environment. The environmental forces can be quantified on the

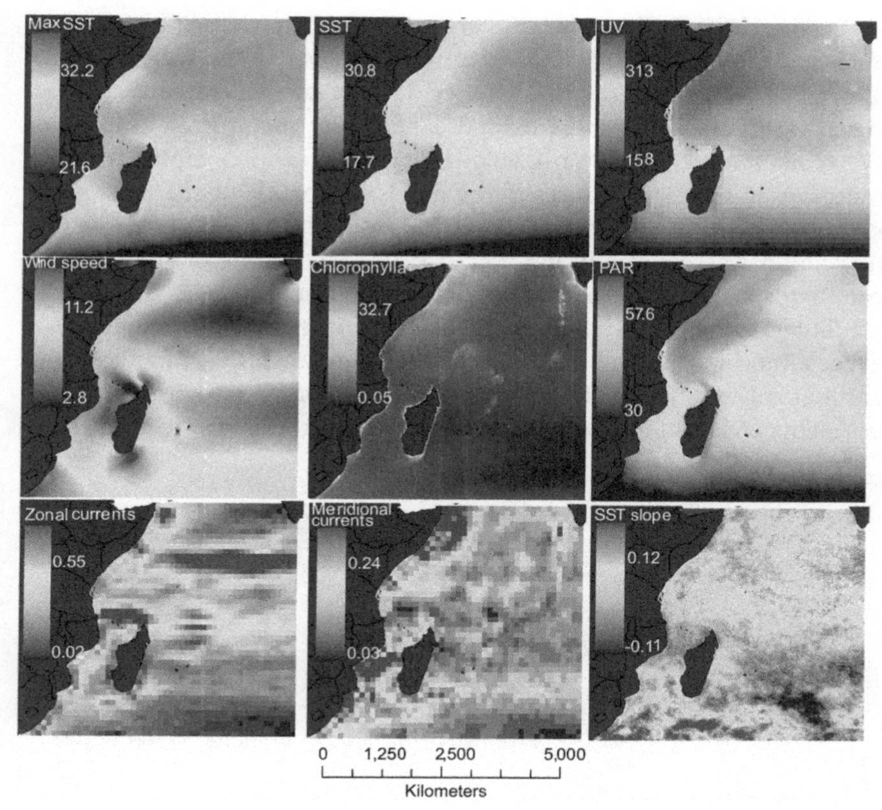

FIGURE 3.7 Distribution of the variation in environmental variables derived from satellite sensor imagines for the western Indian Ocean based on a 4 × 4 km grid for the period 1985 to 2006, except for chlorophyll-a concentrations. MaxSST = mean maximum sea surface temperature, SST = sea surface temperature, UV = ultraviolet radiation, PAR = photosynthetically active radiations.

large scale with satellite information to provide a basis for measuring environmental exposure and the vulnerability of associated organisms. In the chapter that follows we examine the factors that influence the vulnerability of corals and use this information to predict the consequences of climate change on coral reef communities and associated fisheries.

Conclusions

Climate in the region is associated with large-scale processes such as climate oscillations at many temporal frequencies. These processes at different spatial and temporal scales are slowly being uncovered by the use of climate proxies such as coral cores. The time series arising from the various climate proxies are helping to expose the dynamic nature of the Earth in this region. A series of dry and wet rainfall oscillations over the past few millenniums are associated with periods of prosperity and hardship for the people of the region (Verschuren et al. 2000, Wang et al. 2005). There is an emerging warming trajectory in the region reflected in a number of proxies and confirmed by various sources of environmental data. This trend began late in the 19th century and has accelerated in the past 50 years. It is leading to more drought conditions in most of the region, extreme weather, and stressful temperature conditions for marine and other organisms. This trend and heterogeneity is expected to increase and the net effects will challenge the region's ecology and human well-being. The specifics of these impacts, responses, and suggested management interventions are discussed in the following chapters.

4

Climate Change and the Resilience of Coral Reefs

Climate change is predicted to affect coral reefs through three main mechanisms: (1) rising and extreme sea-surface temperatures that results in widespread coral bleaching and mortality, (2) ocean acidification that can potentially inhibit calcification and increase dissolution of the calcium carbonate skeleton of calcifying marine organisms, and (3) an increase in extreme weather events that increase the physical disturbances on coral reefs. However, not all reefs will be equally affected by climate change (Selig et al. 2010). Site-specific environmental properties and the type of ecological assemblage that is present will influence the ways that coral reefs respond to climate-related events such as coral bleaching. Of these three main threats, temperature extremes and coral bleaching are the best studied and are the primary focus of this chapter. It begins with a short introduction to the concepts of adaptation to environmental stress, followed by a review of climate change effects on coral reefs, and the environmental factors that cause coral bleaching. The chapter finishes with a discussion of how to prioritize management based on the exposure of reefs to climate change.

Scales of Adaptation, Resilience, and Stress

In biological terms, an adaptation can be defined as the ability of populations to change in ways that promote their persistence over ecological and evolutionary time. Biologists and ecologists acknowledge three levels of adjustment according to the scale or time taken to make the change. These are, in increasing time for change, (1) acclimatization of the organism, (2) changes in species abundance or community composition, and (3) genetic adaptation or changes in the frequency of genes and associated traits in a population. These three metrics of adjustment and adaptation are important components of ecological resilience (Holling 1973), which is the ability of ecosystems to maintain an ecological state through resistance and recovery from disturbances.

Acclimatization refers to the ability of organisms to change their physiology to allow them to adjust to environmental change. Examples of this include altering metabolism so that key resources such as oxygen, carbon dioxide, inorganic nutrients, and light are used at rates that match the supply from the environment. Acclimatization may require that the appropriate genes be activated to regulate these physiological processes or through epigenetic memory, which is an emerging area of cellular biology that uncovers how cells retain the memory of past changes and can respond quickly when conditions return (Whitelaw and Whitelaw 2006, Rodriguez-Lanetty et al. 2009). Community change refers to changes in the relative abundance of species in an ecosystem, such that the most successful species increase their relative abundance or dominance. Changes in the relative abundance of different types of *Symbiodinium*, unicellular algae that live in coral cells, following a disturbance is discussed later in the chapter (van Oppen et al. 2009). Last, genetic adaptation is the frequency of specific alleles in a population, which change by differential reproduction and survival of individuals. Adaptation can also occur through the emergence of novel alleles that may arise from mutation and benefit individuals with these alleles, but this is rare because most mutations are deleterious.

The time scale of disturbance will influence the ways an organism will adjust. When change is rapid, acclimatization is the most likely response (on the scales of hours to days). Community change is likely on the scale of days to decades, and genetic adaptation occurs on the scale of decades to millenniums, depending on the life span of the species. The capacity for each type of response will depend on the scale of change; the genetic diversity and options; and the turnover of the genes, individuals, and species in the populations and community (Edmunds and Gates 2008).

Environmental Stress

One view of evolutionary history is that major disturbances reset the "evolutionary arms race" that promotes complex adaptations to competition and predation during periods of low environmental disturbance (Vermeij 1987). There is a trade-off between evolutionary fitness under situations of predation and competition versus conditions of environmental stress. Thus, disturbances are predicted to promote species that have fewer adaptations to competition and predation and consequently more tolerance to disturbance and stressful environments. Over ecological time, when species abundance changes after a disturbance, communities can be composed of species with characteristics adapted to low or high environmental stress and disturbance. In general, low disturbance or late-succession communities and ecosystems have been shown to be (1) composed of more species with narrower niches, (2) fewer dominant species, (3) species with greater tolerance to competition and predation, (4) species intolerant to environmental stress or extremes, and (5) more species feeding higher in the food web (Odum 1985, Rapport and Whitford 1999, McClanahan 2002a).

Climate Change and Coral Bleaching

Corals, as described later, are an interesting case of the biological complexity that can arise from the interactions and adaptations among species over evolutionary time. Corals are composed of a coelenterate host that is a carnivore (evolutionarily related to jellyfish) and microscopic dinoflagellate algae living within the host tissue. Within a square centimeter of a host tissue, more than 1 million of these unicellular algae are found (Fig. 4.1). These algae photosynthesize and produce carbohydrates, and most of the organic matter is transferred to the coral host. The rate of calcification is enhanced by this symbiotic relationship. These processes are all threatened by climate change, which can potentially undermine the calcification and productivity processes in these low-nutrient environments.

These unicellular algae are part of what gives corals their color, which can range from yellow to dark brown. These algae, known as zooxanthellae, belong to the genus *Symbiodinium* and were originally classified as a single species—*S. microadriaticum*. The advancement of modern molecular taxonomic techniques has shown, however, that there are in fact hundreds of species of *Symbiodinium* with large differences in their physiology (LaJeunesse et al. 2003, LaJeunesse et al. 2004). Corals can have low or high fidelity with specific *Symbiodinium* species (LaJeunesse et al. 2004, Baker and Romanski 2007). This coral-algae symbiosis is just one of a very large number of species relationships associated with the corals. Other examples include microbes

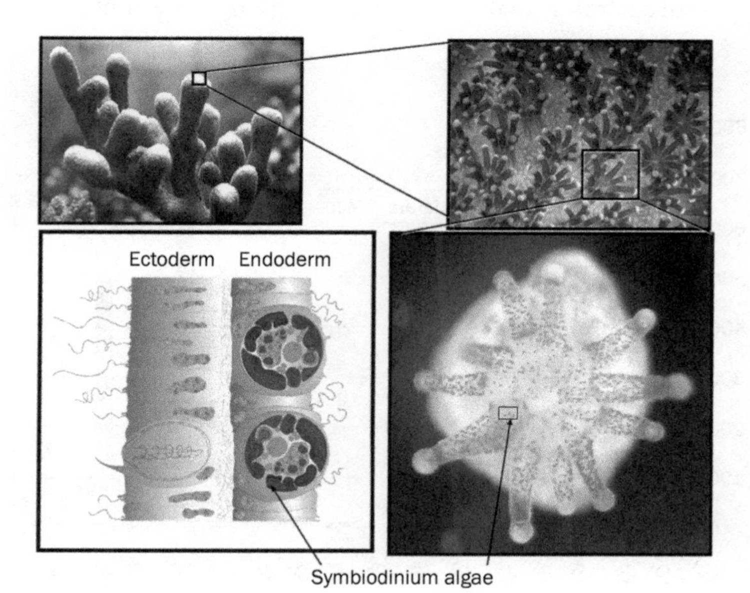

FIGURE 4.1 Diagram of a coral animal and the symbiotic algae it hosts.

Sources: images courtesy of E. Tambutté and A. Moya.

that inhabit the mucus, and skeletons, which include fungi among others (Wegley et al. 2007).

Bleaching is a stress response common among scleractinian and alcyonarian corals, clams, and anemones that causes *Symbiodinium* to leave or be expelled, leaving the animal tissue pale or white (Glynn 1993, 1997). Bleaching is caused by high or low water temperatures, excessive ultraviolet radiation, aerial exposure, high or reduced salinity, high sedimentation, pollutants, or toxins (Coles and Brown 2003). The density of *Symbiodinium* in the coral host tissue changes across seasons and environmental disturbances but "mass-bleaching," in which numerous coral species exhibit unusually high losses in *Symbiodinium*, reflects extreme environmental stresses (Glynn 1991). Recent severe mass-bleaching events have primarily resulted from high sea-surface temperatures linked to global climate change (Oliver et al. 2009; Fig. 4.2). In the most extreme example, the mass-bleaching in 1998 resulted from severe El Niño conditions combined with the Indian Ocean dipole (Saji et al. 1999), which greatly increased sea-surface temperatures throughout the much of the tropics and devastated many of the Earth's coral reefs (Goreau et al. 2000, Wilkinson 2000).

The 1998 global mass-bleaching was the most devastating and widespread bleaching event ever recorded and contributed greatly to increased acceptance of global climate change as both a real phenomenon and a significant threat to entire ecosystems (Walther et al. 2002). The significance of this event was highlighted by the death of coral colonies that had survived 100 to 1,000 years of environmental fluctuations (Hodgson 1999, Mumby et al. 2001).

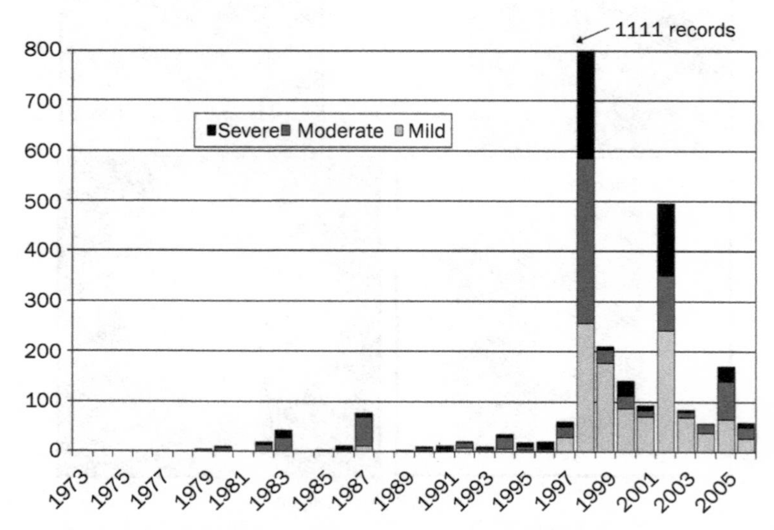

FIGURE 4.2 Reported coral bleaching events based on compiled ReefBase data.
Source: Oliver et al. (2009).

Effects of the 1998 bleaching were spatially variable at both small reef and large geographic scales (Marshall and Baird 2000, McClanahan et al. 2007a, Ateweberhan and McClanahan 2010). Since the 1998 global mass-bleaching event, there have been other instances of large-scale bleaching, particularly in the Caribbean, Australia, and subtropical Indian Oceans (Berkelmans et al. 2004, Donner et al. 2007, McClanahan et al. 2007a). More coral bleaching on the scale of the 1998 event seems inevitable given sustained and ongoing climate change (Donner et al. 2005).

Environmental Factors that Cause Coral Bleaching

Coral bleaching can be caused by a variety of environmental factors because, as noted, it is a generalized response to environmental stress. Rapid, anomalous, and sustained high temperatures over the course of summer do, however, appear to be a key trigger. In the western Indian Ocean region, relationships between environmental variables and the intensity of bleaching were analyzed using historical satellite-derived environmental data and reported observations of bleaching in the Indian Ocean (Fig. 4.3; Maina et al. 2008). Mean and maximum historical temperatures and degree heating weeks (number of weeks that the temperature is at least 1°C above the warm-season baseline) appear to be the main temperature variables that are positively associated with bleaching severity. In contrast, temperature variation (expressed as coefficient of variation or CV) and rates of temperature rise during the past two decades are negatively associated with bleaching. The ways in which each of these relationships influences bleaching will need to be worked out with more field and experimental research, but this preliminary analysis provides a basis for forecasting the conditions when bleaching is most likely to occur. Generally speaking, the conditions of unusually stable warm water, high light, and reduced water motion appear to encourage the severity of bleaching (Eakin et al. 2009).

TEMPERATURE VARIABILITY

It is not only the temperature at the time of the bleaching that is critical but also the usual seasonal temperature range corals are accustomed to (Fig. 4.4; Ateweberhan and McClanahan 2010). Coral mortality tends to be high in locations with low annual variability in sea-surface temperature (SST) and in places with two temperature extremes, both strong winter and summer temperatures. Locations with moderate annual SST variability tend to have lower coral mortality. These locations are the most resistant to these large-scale disturbances and are mostly situated in areas with high water retention, such as the region on leeward sides of islands, such as inner Zanzibar, and

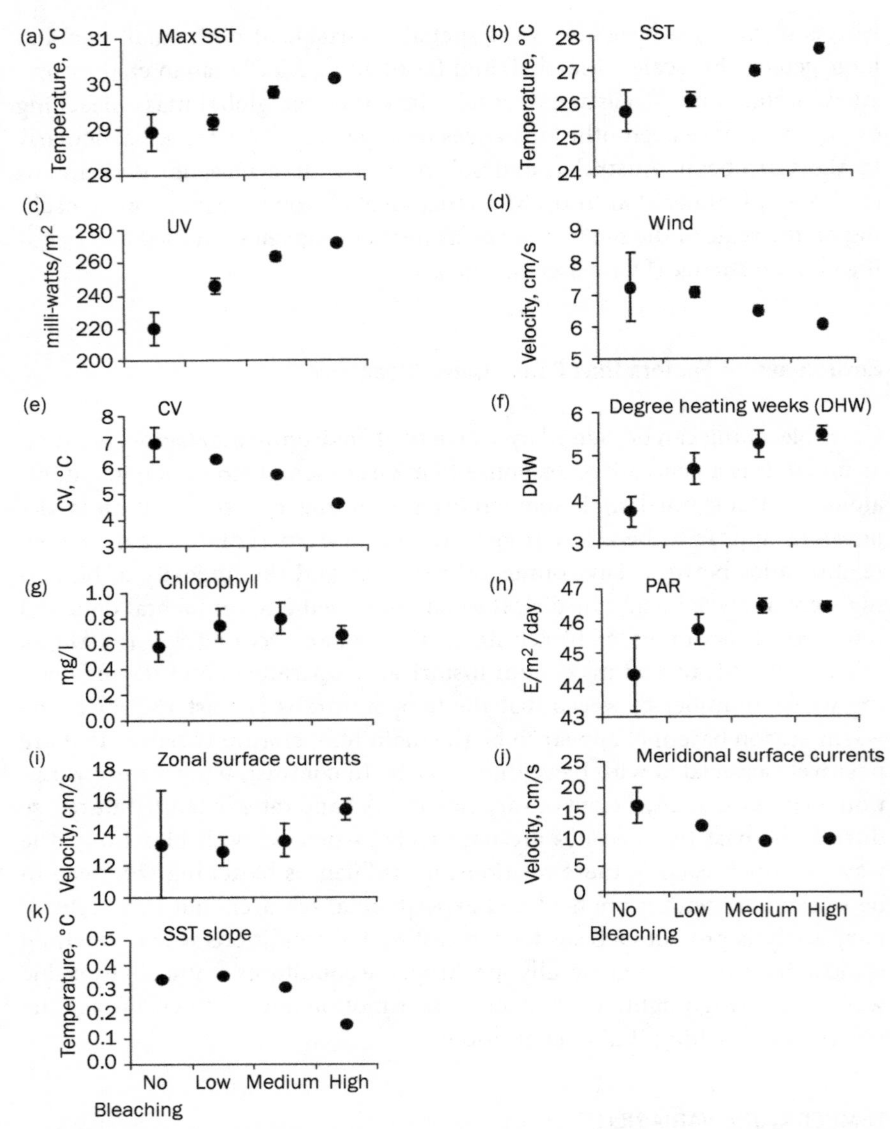

FIGURE 4.3 Relationships between historical environmental factors (7 to 21 years before the bleaching) derived from satellite versus reported severity of bleaching events in the western Indian Ocean. SST = sea surface temperature, UV = ultraviolet light, CV = coefficient of variation in water temperature or a measure of temperature variation, PAR = photosynthetically active radiation, SST slope is the historical rate of rise in water temperature at a site.

Source: Maina and colleagues (2008).

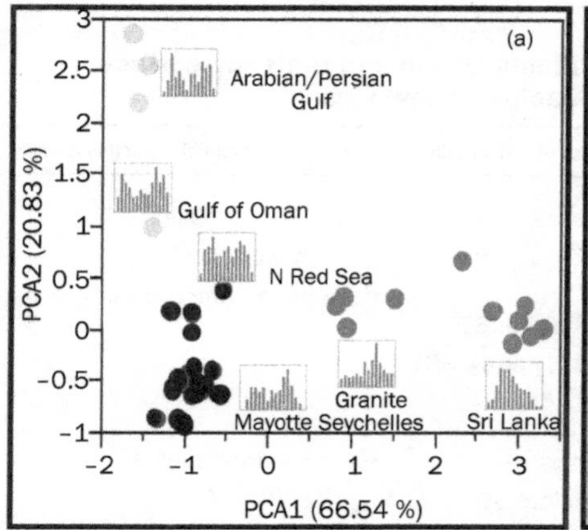

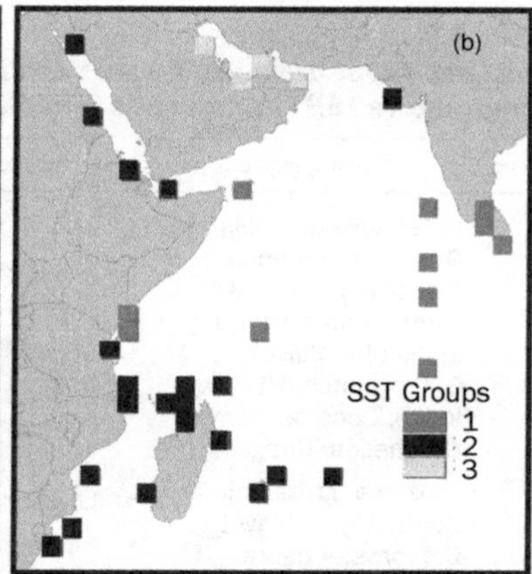

FIGURE 4.4 The three major sea surface temperature groups obtained from a principal component analysis (PCA) (A), and their spatial distribution in the western Indian Ocean (B). The first PC axis (PCA1) separates areas of unimodal from those of bimodal temperature distributions. The second principal component axis (PCA2) separates the bimodal groups into those with large temperature variation (SD) and stronger bimodality and those with weaker to moderate temperature variation and bimodality. This technique separates sites into three groups, which are illustrated by the insets of temperature frequency distribution histograms of representative west Indian Ocean (WIO) reef areas. Group 1 has high mortality and unimodal distributions and includes Sri Lanka (right skewed narrow) and Granite Seychelles (left skewed narrow). Group 2 has low mortality and weakly bimodal distributions and includes Mayotte (left skewed, weak bimodality) and the northern Red Sea (symmetrical, intermediate bimodality). Group 3 has high mortality and strongly bimodal distributions and includes the Gulf of Oman (left skewed, strong bimodality) and the Arabian/Persian Gulf (symmetrical, strong bimodality).
Source: Ateweberhan and McClanahan (2010).

the subtropics, including South Africa and the northern Red Sea. The compilation of seawater temperature and coral mortality by Ateweberhan and McClanahan (2010) identified the areas with moderate temperature variability as least vulnerable (Table 4.1). Western Indian Ocean reefs were generally categorized into three major SST groups of varying coral cover change and SST variability, and these sites may provide a basis for understanding their vulnerability to climate change (Table 4.1). If the condition when these reefs were sampled is taken to represent vulnerability in the region, then we can expect that these patterns will repeat themselves in the future and provide a basis for prioritizing protective management.

OTHER INTERACTIONS

Some environmental disturbances will reduce the effect of another disturbance; others will have no additional effect, while others result in an effect that is larger than the sum of the two disturbances individually. These are

TABLE 4.1

The major reef areas in the western Indian Ocean and their sea surface temperature (SST) properties and bleaching responses

Group	Country/region	Characteristics of SST variation	Environmental and bleaching responses
1	Lakshadweep (India), Gulf of Manar (India-Sri Lanka), Sri Lanka, central and southern atolls of Maldives, Gulf of Kutch (NW India), Socotra, Kenya, Seychelles, Chagos	• Higher kurtosis or peakiness (> –1.0) • Low SD • Socotra higher SD • Unimodal, narrow SST distribution	• High vulnerability • High mortality • Socotra lower mortality
2	Tanzania, Mozambique, South Africa, Mayotte, Comoros, Aldabra, Madagascar, Reunion, Mauritius, Rodrigues, Red Sea, Gulf of Aden, Gulf of Kutch	• Low kurtosis (< –1.0) • Intermediate SD • Low bimodality	• Low to intermediate vulnerability • Low to intermediate mortality • Tanzania moderate to high mortality but good recovery • Aldabra moderate mortality
3	Arabian/Persian Gulf, Gulf of Oman	• Low kurtosis (< –1.2) • Extreme SD (≥ 2) • Strong bimodality and low uniformity	• High vulnerability • High mortality

Note: SD = temperature variation standard deviation.
Source: Ateweberhan and McClanahan (2010).

referred to as acclimatizing, neutral, and sensitizing stresses and they may be equally common in nature (Darling and Cote 2008). The bleaching response is influenced by other factors, such as light, winds, water flow, fishing, and pollution. One study in the Caribbean found, for example, that corals exposed to warm water temperature anomalies, fishing, and pollution had slowed their growth rates compared to those not exposed to these anthropogenic stresses (Carilli et al. 2010). Light becomes a stress when temperatures are high (Baird et al. 2009); therefore, photosynthetically active radiation (PAR) and ultraviolet (UV) light are positively associated with bleaching severity in the WIO region (Coles and Brown 2003, Maina et al. 2008).

Winds and currents also play a key role in bleaching. Coral bleaching has been most severe in areas when the winds drop and currents run meridionally (north–south). The influence of current variability on coral mortality has been observed at a range of scales. Along a 1,000-km stretch of the east African coastline, reefs that experienced low temperature variation in the north suffered higher relative mortality than reefs in the south where temperature variation was higher (McClanahan et al. 2007c). Smaller scale examples from Mauritius show that in 2005, the island's leeward reefs bleached less than their windward reefs (McClanahan et al. 2005a). Likewise, across the 1998 event, inner lagoons around Chagos Island had lower mortality than outer reefs (Sheppard et al. 2002). In Kenya, low water flow and high temperature variation lagoons fared better than high water flow and low temperature variation lagoons (McClanahan and Maina 2003). Consequently, the current state of observations suggests an important role for acclimatization in reducing the effects of temperature anomalies.

In general, high water flow can be beneficial to coral because it delivers and flushes essential nutrients and dissolved gases (Nakamura and van Woesik 2001). During bleaching events, however, wind-driven flow is often reduced, which creates an additional and unaccustomed stress. For example, the frequency of a coral disease was found to be greater in reef lagoon environments in Kenya with high compared to low water flow. Higher bleaching in these environments may also make them more susceptible to diseases (McClanahan et al. 2009a). The low current and associated high temperature variation create some acclimatization and adjustment or adaptation that reduce the stress of periodic or rare extreme warm temperature anomalies, such as El Niño and Indian Ocean dipole events.

Impacts of Coral Bleaching

CHANGES IN CORAL COVER

Coral cover change has been used as the main metric for investigating impacts of climate and other disturbances on coral reefs. Several studies have indicated significant changes in cover due to climate change, the 1983, 1987, and 1998 El Niño southern oscillations (ENSO) being the globally prominent climatic events. The events of 2002 and 2005 also had strong effects on the Great Barrier Reef and the Caribbean, respectively. The 1998 event caused a high and global coral mortality, the WIO region suffering one of the highest mortalities because the ENSO combined with the Indian Ocean dipole (IOD) to create an unusually strong warm anomaly (Fig. 3.2).

Despite the widespread nature and strong impact of the 1998 event, WIO wide responses in bleaching mortality and change in coral cover varied by region and among reefs within a region. A compilation of change in coral cover based on 36 locations where before and after coral cover could be estimated, showed that the 1998 bleaching events caused an overall decline of about 40% coral cover across the western Indian Ocean (Ateweberhan and McClanahan 2010). The highest mortality generally occurred in the central and northern regions, with the exception of the Red Sea and Gulf of Aden (Fig. 4.5). The most severely affected reef areas were in southern India, Sri Lanka, and central atolls of the Maldives and Granite Seychelles. The Red Sea, Mayotte, Comoros, southern Mozambique, South Africa, Madagascar, Reunion, Mauritius, and Rodrigues were the least affected. Coral cover losses were associated with changes in the coral species composition, reef complexity, erosion of coral skeletons, and the fish fauna.

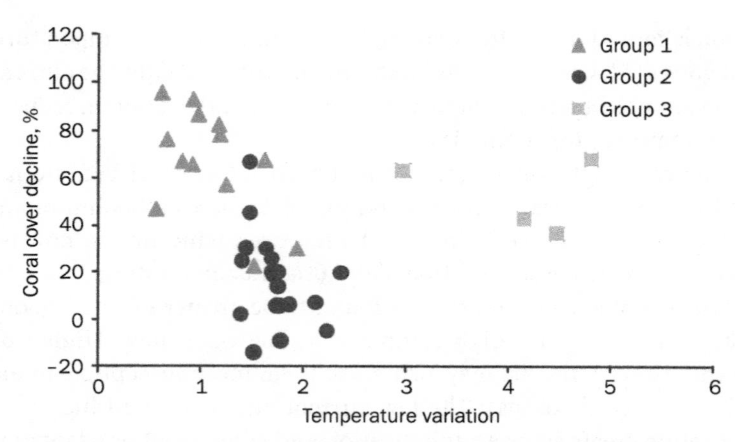

FIGURE 4.5 Relative decline in coral cover (%) in the western Indian Ocean as a function of sea surface temperature variability measures.

Source: Ateweberhan and McClanahan (2010).

CHANGES IN CORAL COMMUNITIES AND DIVERSITY

On coral reefs, all species hosting the symbiotic zooxanthellae (i.e., Cnidaria, Foraminifera, Porifera, and Mollusca), calcifying organisms (i.e., Scleractinia, Corallinaceae, and Halimedaceae), and their obligate symbionts are susceptible to bleaching (Glynn 2000). All of these groups are susceptible to bleaching and mortality during warm anomalies but have differential responses. Even different species of coral vary in their susceptibility to bleaching. These differences can result in changes to coral community composition and diversity after bleaching events. Thus, the history of disturbance on a reef can be evaluated by the relative abundance of susceptible and robust taxa currently on the reef.

Branching and plating coral species, such as *Acropora* and *Montipora,* generally dominate undisturbed reefs but are very susceptible to the effects of bleaching. The relative abundance of these two coral species in the WIO declined in direct proportion to the thermal stress during the 1998 coral bleaching event (McClanahan et al. 2007a). Massive and submassive corals, such as *Porites* and species in the Faviidae, have been found to be more robust to bleaching and tend to dominate reefs that have been disturbed (Fig. 4.6). Additional changes to community composition can occur when the reduction of competitively dominant species, such as *Acropora* and *Montipora*, releases subordinate and stress-tolerant species from competition.

The 1998 bleaching event resulted in substantial changes in the composition of coral communities in most of the northern Indian Ocean (McClanahan et al. 2007a). In contrast, the southern Indian Ocean communities have maintained more of the branching and encrusting forms that typify undisturbed reefs, with the exception of Reunion Island. The transition between these two reef states is at the international border between Kenya and Tanzania.

FIGURE 4.6 Photos of the species that are highly susceptible to bleaching: (a) *Acropora* and (b) *Montipora;* those that are less susceptible include (c) massive *Porites* and (d) *Favia*.

Climate events such as bleaching may affect the diversity of reefs through change in community structure but also by causing local extinctions of taxa. The taxa that are prone to local extirpation and possible extinction include those that (1) have narrow environmental limits; (2) have restricted ranges; (3) have small population sizes, often caused by restricted habitat requirements or being high in the food web, thus limited by resources; (4) predominantly reproduce asexually and have low dispersal capabilities; (5) grow slowly; and (6) are endemic to biogeographic regions with small continental shelves (Roberts and Hawkins 1999, Hughes et al. 2002). Species losses at the community level do not necessarily lead to extinctions when there is sufficient spatial and temporal heterogeneity in the environment (McClanahan and Maina 2003). Heterogeneity in space and time is often accompanied by heterogeneous distributions of species and populations in space, which is known as the meta-population nature of species distributions. The patch distribution decreases the probability that sites and regions will have a uniform response to a disturbance allowing some populations to persist (McClanahan 2002a).

Loss of coral species can be partially predicted by knowing three important aspects of the taxa's life history: (1) their sensitivity to a disturbance;

(2) abundance; and (3) geographic distribution. Local loss and extinction is higher for species with high sensitivity, low abundance, and small geographic distributions (Figs. 4.6 and 4.7, McClanahan et al. 2007a). Some of the most temperature-sensitive taxa such as *Montipora*, *Acropora*, and *Pocillopora* are widely distributed and likely to persist over climate change disturbances, even if their numbers are reduced considerably and locally extirpated in some sites. Taxa such as *Plerogyra*, *Oxypora*, *Plesiastrea*, *Gyrosmillia* and *Physogyra* that have high bleaching sensitivity and limited distributions could be prone to local extinction. The vulnerable genera only have one or two species per genus in the WIO and these indicate that these genera could go regionally extinct with increasing climate change disturbances.

The extent of local losses and even extinctions may be greatly influenced by the spatial distribution of the coral taxa. Surviving populations are more likely to be reseeded by larvae from nearby and relatively unaffected populations (Ayre and Hughes 2004). Consequently, isolated reefs are expected to be more sensitive to declines in local populations, whereas well-connected reefs, such as those along continental margins or large archipelagos are hypothesized to have better chances of recovery. Recovery is expected to be faster if larger adults survive a bleaching event. Mature corals are the most reliable source of new recruits, and growth of surviving corals leads to more rapid increases in coral cover compared to larval settlement and growth of

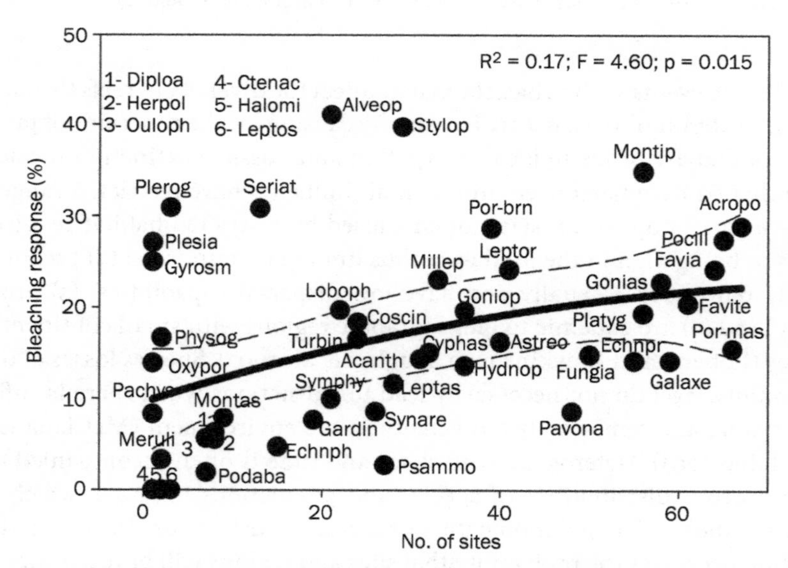

FIGURE 4.7 Plot of the bleaching sensitivity of the common hard coral taxa in the Indian Ocean versus the number of sites where they were found, which is an indication of their relative abundance in the region. The relationship between sensitivity and abundance is statistically significant. The first six letters of the hard coral genera names are provided.

Source: McClanahan et al. (2007a).

new individuals (Connell 1997, Hughes et al. 2000, Baird and Marshall 2002). However, large colonies are also more susceptible to bleaching than small colonies and recruits (Loya et al. 2001).

A study of coral diversity with comparable data before and after the 1998 bleaching event found that immediately after this disturbance there was a temporary reduction in the number of coral taxa in Kenya, but this loss recovered in a few years (McClanahan et al. 2005b). Another more detailed study in Kenya found that the most diverse reefs lost the most coral taxa, which indicates that stable environmental conditions may promote species richness but could also contain more vulnerable taxa (McClanahan and Maina 2003). This indicates a trade-off between species richness and exposure to bleaching. The most resilient reefs have the least number and are less likely to lose more taxa because they have been previously disturbed and probably have some community-level pre-adjustment or adaptation to disturbance (Darling and Cote 2008, McClanahan 2008). The perspective that high diversity reefs are more resistant to disturbances is not supported by this empirical field study; more work is needed to determine how general this pattern is and whether recovery rates are faster on high diversity reefs after disturbances.

At the scale of the whole Indian Ocean, there was not a significant relationship between taxonomic richness and absolute or relative change in coral across the 1998 event, with sites scattered in all regions of the plot (Ateweberhan and McClanahan unpublished data; Fig. 4.8). High-diversity areas that suffered high mortality were located mostly in the central-northern WIO, while high- diversity reefs in the triangle between Tanzania, Mozambique, and Madagascar suffered the least. The results indicate that factors that control regional species richness and climate change responses, measured by coral loss, are probably different.

OTHER CHANGES TO THE BENTHOS

The impacts of climate change on coral reefs have not been limited to hard corals, but more is known about corals because as the main foundation species, they generate the most scientific and monitoring focus. Less is known about the responses of other major benthic groups to climate disturbances, including soft corals, sponges, and different algal species and forms—crustose coralline, green calcareous, turf, and fleshy algae. For example, soft corals and green calcareous algae declined immediately after the 1998 temperature anomalies and slowly recovered (McClanahan 2008). Calcareous green algae and red algae do, however, appear to benefit in the medium term from these disturbances because they are able to quickly colonize the vacant space created by coral mortality and they have high grazing tolerance due to their calcified thallus and chemical defenses. Soft corals are slower to respond, and the long-term consequences for them are unclear. In the better studied reefs,

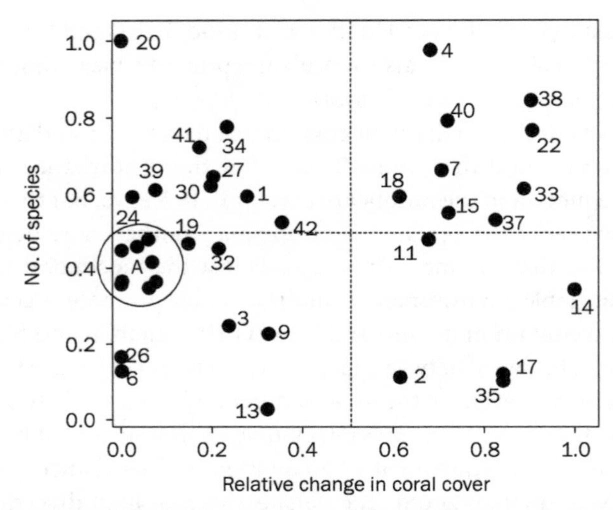

FIGURE 4.8 Scatterplot of coral diversity-mortality relationship in the western Indian Ocean. Values are standardized from 0 to 1 relative to the lowest and highest proportion of coral cover change, respectively. High and low areas are separated at a conventional 50% cut-off level (dashed lines). Sites are 1. Aldabra, Cosmoledo, Faarquhar, 2. Arabian Gulf, 3. Arabian Sea, 4. Australia NW, 5. Australia SW, 6. Bangladesh, 7. Chagos, 8. Cocos Keeling, 9. Comoros, 10. Djibouti (Gulf of Aden), 11. Gulf of Oman, 12. India (Andaman-Nicobar), 13. India (Gulf of Kutch), 14. India (Gulf of Manar), 15. India (Lakshadweep), 16. India (West coast patches), 17. Kenya-N, 18. Kenya-S, 19. Madagascar-NE, 20. Madagascar-NW, 21. Madagascar-SW, 22. Maldives, 23. Mauritius, 24. Mayotte, 25. Mozambique-S (Pemba), 26. Myanmar (Burma), 27. Red Sea-C, 28. Red Sea-N, 29. Red Sea-S, 30. Reunion, 31. Rodrigues, 32. Seychelles, 33. Socotra, 34. Somalia, 35. South Africa, 36. Sri Lanka, 37. Tanzania-C (Mafia), 38. Tanzania-C (SongoSongo), 39. Tanzania-N (Zanzibar), 40. Tanzania-S (Mnazi Bay), 41. Thailand-Mergui Archipelago. Numbers not shown are contained in the circle marked A.
Source: Ateweberhan and McClanahan (unpublished data).

the immediate response was generally an increase in various types of algae, ranging from small and fast-colonizing turf to erect fleshy brown to slower colonizing calcifying red and green algae (McClanahan et al. 2001). In some cases, these changes are short term lasting less than a few years, while in other conditions they appear to persist for many years and may represent a permanent shift to an erect-algal dominated ecosystem (Hughes 1994). The consequences of this change to fishes are discussed in more detail in the following chapter.

EROSION OF CARBONATES

Calcifying organisms, especially corals, build the reef structure that is critical habitat for fish and other invertebrates. This structure is, however, constantly being eroded by organisms that bore away the calcium carbonate structure in a process called bioerosion (Glynn 1997). The key organisms driving bioerosion are parrotfish, sea urchins, sponges, mollusks, microalgae, and various microbes that live in the coral's carbonate skeletons. Maintaining a reef structure will become increasingly challenging with climate change as coral bleaching exposes carbonate surfaces to increased bioerosion, and calcification rates are reduced through coral mortality. Even at sub-lethal levels, coral bleaching reduces calcification rates (Suzuki et al. 2003, Rodrigues and Grottoli 2006). Reduction in the calcification rates as the result of change in ocean chemistry may result in the formation of less dense skeletons, which are highly susceptible to rapid physico-chemical and biological erosion (Tribollet et al. 2002, Carricart-Ganivet 2007).

Parrotfish tend to be the dominant carbonate eroders when they have not been removed by fishing (Carreiro-Silva and McClanahan 2001). Sea urchins benefit from overfishing and can be responsible for considerable carbonate erosion on heavily fished reefs (McClanahan 2000a). Sponges and microbes are less important eroders and can be influenced by pollution and water quality issues such as dissolved organic nutrients and plankton concentrations. Overfishing is often accompanied by other stresses such as pollution that induce coral and sea urchin diseases favoring algal dominance and bioerosion by sponges and microbes as witnessed in the Caribbean (Hughes 1994). Heavy fishing and pollution can influence the bioerosion rates, and these stressors should be reduced, especially during and immediately after warming events to avoid losing reef structure (Carreiro-Silva and McClanahan 2001).

Other Climate Change Impacts on Coral Reefs

OCEAN ACIDIFICATION

The increase in atmospheric CO_2 has resulted in significant changes in the oceanic carbonate chemistry that is expected to reduce the ocean's bicarbonate ions, aragonite saturation, and pH. Of particular interest to coral reefs is the reduction in bicarbonate ions, which has been shown to decrease calcification rates (Jury et al. 2010b). Aragonite saturation and pH are also often associated with changing calcification, but coral responses from experiments have been variable, likely due to the difficulties of separating these three ocean chemistry variables (Langdon and Atkinson 2005, Jury et al. 2010b).

The global oceans now have a seawater carbonate ion concentration that is lower by 30 μmol/kg than the pre-industrial level and are more acidic by 0.1-pH unit (Dore et al. 2009). Current estimates based on aragonite saturation and temperatures are that for the very likely CO_2 atmospheric concentration of 450 parts per million (ppm) and 2°C warmer, coral reef organisms will exhibit very low calcification rates. Silverman and colleagues (2009) suggested that reefs will cease to grow and start to dissolve at 560 ppm CO_2 or about 3°C warmer. This is a larger predicted impact than previous estimations of a 40% reduction in calcification at 560 ppm CO_2 (Kleypas et al. 2006). Nevertheless, because aragonite saturation is more easily estimated in the future and on a global level than bicarbonate concentrations, which appears to be largely responsible for calcification, there is considerable more research needed to determine the potential effects of the changing ocean carbonate chemistry and its variability on calcification.

Recent reports have documented a decline in the mean annual calcification rates of massive *Porites* and *Diploastrea* by up to 30%–40% in the later years of the 20th century, associated with rising seawater temperatures and increased bleaching events (Cooper et al. 2008, De'ath et al. 2009, Cantin et al. 2010). This is expected to reduce coral survival, and reduced growth could undermine the competitive ability of corals and other calcifying organisms. The strongest and earliest effects will be felt at higher latitudes, but they will eventually reach equatorial regions and will be at dangerously low levels 50 years from now if the current carbon emission rates continue. Acidification is not locally manageable, as it requires reductions in carbon dioxide emissions and uptake by the ocean.

There may be ways to manage coral reefs that reduce acidification impacts on corals. For example, photosynthesis by seagrass changes the near shore carbonate chemistry and could acclimate and buffer corals to changes in oceanic acidification (Semesi et al. 2009). Moreover, a review of studies found that corals receiving supplemental nutrition calcified at normal rates even when exposed to elevated CO_2 conditions (Cohen and Holcomb 2009). Consequently, these findings suggest some patchy responses to elevated CO_2, depending on local conditions for nutrients and seagrass. Most predictions of climate change suggest that on the global scale climate change will reduce oceanographic nutrients and productivity. In the Indian Ocean, productivity changes have been measured and are decreasing in the southern but increasing in northern Indian Ocean (Gregg et al. 2003). Therefore, the buffering effect of high nutrients on coral growth is unlikely to have broad-scale effects, and high nutrients can also have a number of negative effects on coral reefs (Fabricius 2005). Nevertheless, efforts to protect and promote seagrass beds need to be examined further as a way to reduce the negative effects of changing oceanic carbonate chemistry.

CYCLONES

The intensity of tropical cyclones is expected to increase with global warming (Trenberth 2007). Tropical cyclones can cause considerable damage to the reefs via strong wave action that causes physical destruction of colonies, decreasing salinity from floodwaters, and increased freshwater runoff. In the

Caribbean where hurricanes are a frequent event, coral cover is, on average, reduced by about 17% immediately following strong hurricanes (Gardner et al. 2005). Webster and colleagues (2005) suggest that global satellite data of tropical cyclones indicates a 30-year trend toward more frequent and intense hurricanes, particularly a large increase in the number and proportion of hurricanes reaching categories 4 and 5. Model predictions also show that at the current rate of global temperature rise, nearly a doubling of the frequency of category 4 and 5 storms is expected by the end of the 21st century (Bender et al. 2010). In the western Indian Ocean, there is evidence for stronger but not more frequent cyclones (Kuleshov et al. 2008). Cyclone effects are often short-lived unless associated with large terrestrial sediment inputs and they are not expected to pose threats comparable to those of coral bleaching and reduced calcification and carbonate dissolution.

Despite their destructive effects, there is also evidence that some hurricanes temporarily reduce thermal stress on corals by drawing away heat and mixing cooler subsurface waters with the warmer surface waters. Manzello and colleagues (2007) suggested that hurricane-induced cooling was responsible for the differences in the extent and recovery time of coral bleaching between the Florida Reef Tract and the U.S. Virgin Islands during the Caribbean-wide 2005 bleaching event. The cyclone effect was suggested for the weak bleaching reported in Mauritius in 1998 (Turner et al. 2000) and might potentially reduce bleaching effects across the WIO cyclone belt (from Rodrigues to eastern Madagascar). The overall effects of cyclones are probably negative, however, as a compilation of studies in the Caribbean found that cyclones had tripled the rate of the regional decline of these reefs over the past 30 years (Gardner et al. 2005).

Recovery of Corals

The recovery of coral after small-scale disturbances is expected on the scale of 5 to 15 years and at rates of 1% to 10% per year but there are many examples of reefs that do not recover on these time scales after disturbances (Baker et al. 2008). For example, large meta-analyses of empirical field studies of the Indo-Pacific and Caribbean reefs indicate that coral cover is declining at around 0.72% and 5.5% per year over a 30-year period, respectively (Gardner et al. 2003, Bruno and Selig 2007). The fastest recovery is usually due to a few taxa in the Acroporidae and Montiporidae. Growth rates for these species can exceed 10% per year at their peak, which is usually a few years after a disturbance (Thompson and Dolman 2009). When other taxa are dominant, the rates are considerably slower; for example, a long-term study of recovery in Kenya, after 1998, found that rates were closer to 2% per year and few reefs had fully recovered their coral cover 10 years after the disturbance (Darling et al. 2009).

Because warm water can kill the whole colony, there is considerable potential for temperature anomalies to jeopardize the conditions for fast recovery depending on the corals' mode of reproduction and recovery. Corals reproduce in two ways, sexually and asexually. Sexual reproduction creates larvae that survive in the water column for a few hours to six months and disperse distances of tens of meters to thousands of kilometers (Graham et al. 2008a). Asexual reproduction is achieved when a portion of the adult colony breaks off and disperses several meters away before reattaching to the bottom and growing. In some cases, colonies can die back from disturbances but remnant living tissue is left at the base of the coral and this can begin regrowing after the disturbance. In some cases, when bleaching is not severe, portions of the colony hidden from strong light can survive, and these hidden remnants can be the basis for fast recovery. For example, an investigation in the Great Barrier Reef found recovery after a few years as mortality was partial and corals recovered from remnant tissue at the base of the skeleton—the coral equivalent of sprouting from roots (Diaz-Pulido et al. 2009). Recovery will be considerably slower when it is dependent on larval recruitment that requires a nearby undisturbed population and appropriate substratum for settlement (Underwood et al. 2009).

Studies of larval recruitment after bleaching-induced mortality find recruitment to be either lower or considerably variable in space (Glynn et al. 2000, McClanahan 2000b, Guzman and Cortes 2007). Moreover, even high recruitment does not always correspond to fast recovery because other ecological limitations can prevent larval success (McClanahan et al. 2005b, Stobart et al. 2005, Golbuu et al. 2007). There are also examples and indirect evidence that even when coral cover does recover rapidly, as seen in a number of WIO reefs (Baker et al. 2008), the community can shift considerably—toward a mix of opportunistic taxa, such as *Pocillopora*, and robust taxa, such as massive *Porites* and bleaching-resistant faviids (Berumen and Pratchett 2006, McClanahan 2008). The Chagos, Maldives, and parts of Tanzania have, however, recovered to the original dominance of fast-growing species, such as *Acropora* (Sheppard et al. 2002, Wallace and Zahir 2007, McClanahan et al. 2009b).

Which Reefs Do We Focus Protection On?

Information about differing levels of exposure to stress can be used to help inform reef management, but there are divergent perspectives as to whether the focal points for management should be the areas with the highest or the lowest levels of exposure to stress. Assumptions about ecological organization, human values, and purposes of protection underlie the management choices, and these will depend on whether the stated purpose is to preserve undisturbed ecosystems or to promote some other social or economic value, including tourism, fisheries production, or national heritage. For the discussion here, we will assume that the purpose is to preserve an intact but threatened ecosystem.

Advocates of ecological preservation suggest that regions with low environmental exposure to climate change, which are least likely to be exposed to bleaching and coral mortality, are refugia and should be a high priority for conservation (Sanderson et al. 2002, West and Salm 2003). The rationale behind this view is that protecting undisturbed areas will ensure that other anthropogenic local stresses are minimized and some ecosystems and species will remain intact. These intact areas may eventually have the potential to be sources for recolonization and recovery of disturbed ecosystems. These assumptions may be true if the focus is on saving a single site, but the strategy is more complex if the target is protection of a network of reefs and where other considerations, such as total biodiversity, ecological representativeness and uniqueness, and costs of protection, are also considerations (Game et al. 2008a). Game and colleagues (2008b) argue based on model outputs that a risk-spreading behavior where a variety of ecosystem and vulnerability types are protected is likely to have benefits to the total seascape depending on the response of the reefs to disturbance.

Simple models have been developed to evaluate these considerations. One model found for a small increase in costs that the chances of persistence increase considerably. In a specific example of the Great Barrier Reef, a 2% increase in overall costs improved the long-run performance of the reserves by 60%. Based on this model, Game and colleagues (2008b) found that the rate of recovery is critical to deciding whether to protect high- or low-risk sites. Protection of high-risk sites is beneficial when they have high rates of recovery but not when these rates are low. If protected sites spend most of their time in a degraded state, then protection of low-risk sites is recommended. An additional modeling study evaluated the best approaches to protecting coral reefs using coral and symbiont diversity, ecological vulnerability, and the connectivity of reefs and concluded that protecting diverse and highly connected coral communities was most critical for persistence, followed by protecting communities with more thermally tolerant coral and symbiont species (Baskett et al. 2010). These studies provide guidelines for management decisions and indicate the importance of environmental and ecological context once the social values have been determined.

An alternative view is that protection would be best directed at reefs with high exposure to climate change-related stresses. The argument is that these reefs have low numbers of species and ecological redundancy; therefore, extreme climatic events combined with fishing and other disturbances may push these ecosystems "over the edge" more easily than reefs not exposed to stresses or with greater numbers of species and ecological redundancy. Protective management may prevent these systems from shifting to less desirable states, such as sea urchin barrens or algal-dominated systems, where ecological services to people are reduced. Field studies, however, are not supportive of this theory because these ecosystems are typically composed of highly resilient species that are tolerant of a variety of disturbances (McClanahan and Maina 2003, Darling et al. 2009). Essentially, the more vulnerable taxa are likely to have already been removed from the system as a result of past disturbances. Furthermore, there is little evidence that protection from fishing reduced the initial damage or resistance to the 1998 coral-bleaching event in the western Indian Ocean (Graham et al. 2008b, McClanahan 2008).

The evidence for faster recovery in better managed reefs is equivocal (Graham et al. 2008b, Selig and Bruno 2010), and where recovery of cover did occur, the species composition responsible for the cover differed considerably after the disturbance (McClanahan 2008). Remote reef systems, such as the Chagos Islands, have shown rapid recovery after 1998 (Sheppard et al. 2008), but these pristine reefs may be a poor analogue for smaller protected area closures more typical of the management used in tropical countries.

Conclusions

Climate change is reducing the dominance of calcifying coral species that provide the important reef architecture needed to support fish and fisheries and other ecosystem services, including shoreline protection. Reefs are expected to change toward dominance by either slower growing and stress-resistant corals, moderately fast-growing calcifying algae, or fast-growing non-calcifying algae, or other groups such as various soft corals, depending on the level of other stress factors (Norstrom et al. 2009). Climate change disturbances are variable in space and time and associated with the spatial heterogeneity in the many potentially stressful environmental variables. This suggests that the impacts of climate change will not be simultaneous and uniform and this variability will provide time to implement management and human adaptations that can reduce detrimental consequences for people. The connections to fish, people, and management is discussed and synthesized in subsequent chapters.

5

Climate Change and Coral Reef Fishes and Fisheries

Climate change is expected to have both direct and indirect effects on coral reef fishes, including commercially important fisheries. Direct effects include sub-lethal effects of temperature and ocean acidification that can influence physiological processes and reduce reproductive output and change larval supply, movement, and migration. Indirect effects include the loss of living coral tissue and reef structure through the erosion of the exposed coral skeletons, which influence feeding and settlement of fish larvae and ecological interactions. This chapter discusses these issues in more detail. First, we provide a brief summary of the scientific literature on the effects of climate change and coral mortality on fish populations and communities, including a region-wide case study that evaluated the 1998 impacts across the western Indian Ocean (WIO). We then discuss the impacts of climate change on reef fisheries. Based on the information presented here and in the previous chapter, the ecological axis of the adaptive management framework (Fig. 1.1) is developed in Chapter 8 and recommendations for management are described in Chapter 10.

The Impacts of Climate Change on Reef Fishes

LOSS OF CORAL TISSUE

Many species of fishes are strongly associated with coral reefs. Both living coral and the reef structure or topographic complexity play a role in this association. The relative importance of these features depends in a large part on the specific fish and their feeding and habitat needs (Wilson et al. 2006). For example, living coral is important for specialist fishes that depend on corals for food or shelter (Pratchett et al. 2008). Current estimates of the proportion of coral reef fishes with apparent and direct reliance on live corals are 9% to 11% of the total numbers of coral reef species (Jones et al. 2004, Munday et al. 2008).

The role of living coral in structuring reef fish assemblages has been explored by examining the effects of disturbances that kill corals without affecting their skeletal structure, such as coral bleaching, the coral eating Crown-of-Thorns starfish outbreaks, and coral diseases. While these disturbances can have severe effects on corals, they appear to affect only fishes that are highly dependent on corals for settlement, food, and shelter (Fig. 5.1; Pratchett et al. 2008). As an example, one study in which 15% of the local fish species relied directly on corals for food or habitat found that 75% of coral reef fish species exhibited declines in their numbers, and diversity declined by 22%, following a 90% loss in coral cover (Jones et al. 2004). The authors attributed this larger than expected decline to settlement requirements, whereby many species require live coral tissue for successful settlement.

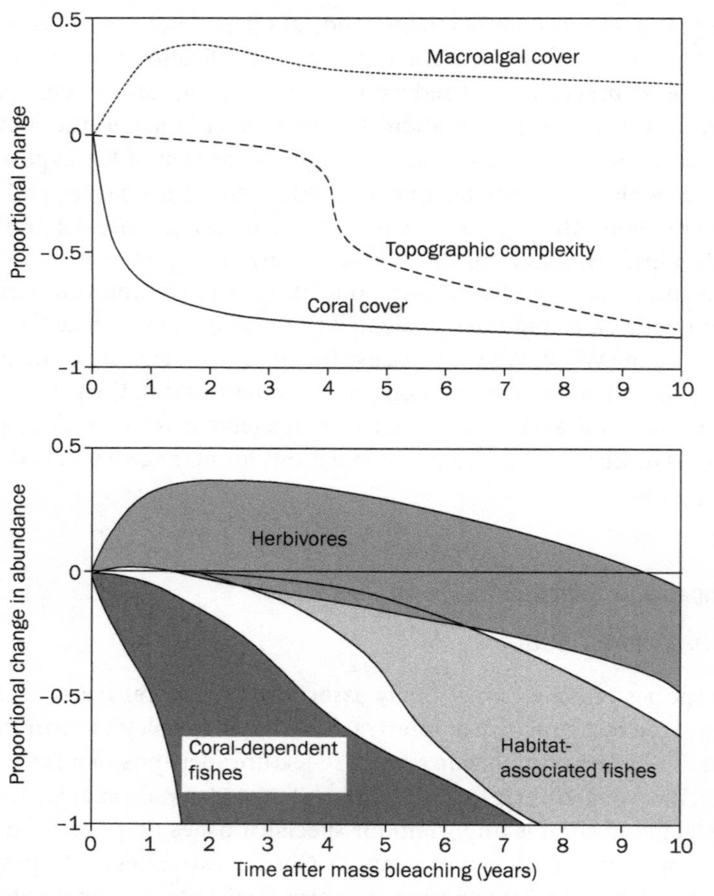

FIGURE 5.1 Proposed responses of the benthic and fish populations to a hypothetical coral mortality event. Coral cover is lost in the first year or two followed by a loss of reef topographic complexity. This leads to different responses for the fish, where coral-dependent fish populations are the first to decline, followed by those that require the complex habitat, and finally a loss of herbivorous fishes.

Source: Pratchett et al. (2008).

LOSS OF TOPOGRAPHIC COMPLEXITY

Most coral reef fishes do not depend directly on living coral tissue but are dependent on the topographic complexity created by their skeletons, although the two are often hard to separate (McClanahan and Mutere 1994). Topographic complexity plays a key role for many reef fishes by providing shelter from predators and creating spatial niches that enhance their diversity (Roberts and Ormond 1987, Hixon and Menge 1991). As discussed in Chapter 4, coral mortality can eventually result in a collapse of the reef structure itself, reducing the diversity of habitats and number of hiding places. Fast-growing branching and plating corals are more susceptible to mechanical and bleaching disturbances than are mound corals, which provide less refuge for coral-dwelling fish (Munday et al. 2008).

The importance of topographic complexity to reef fishes has been examined by studying fish assemblages before and after physical disturbances, such as severe tropical storms and tsunamis. These studies have found large changes in the fish communities, but separating the effects of coral tissue and topographic complexity is challenging (Letourneur et al. 1993). Butterflyfishes feed most often on branching and plating species and have proved useful for separating the two effects because some species are generalists that do not rely on coral, some are obligate specialists that cannot survive without coral, and other species feed facultatively on corals. As predicted, the greatest losses in numbers of butterflyfishes after bleaching and tissue loss were among the obligate specialists, followed by generalist species, while the numbers of facultative species did not decline unless the reef complexity was lost (Graham 2007, Graham et al. 2009). Further, for other species whose juvenile stages need the refuge provided by coral or whose adults live in coral, there is a lag between coral loss and their declines, which is most likely associated with lost complexity (Graham et al. 2007).

LARVAL SUPPLY

The survival of fish larvae is influenced by several mechanisms that will change with rapid climate warming. For example, climate warming is expected to reduce the density of the ocean's surface, increase vertical stratification, change surface mixing, slightly increase global primary production, and change the dominant phytoplankton. Further, the intensification of hydrological cycles is expected to lead to increased terrestrial runoff, river discharge rates, and flooding. These effects on coral reef fishes will be mixed, as small increases in ocean temperature and productivity might enhance the number of larvae surviving the pelagic phase but larger increases or changes in the plankton communities are likely to reduce reproductive output and increase larval mortality (Munday et al. 2009a). Changes to ocean currents could alter larval supply but the responses are likely to vary greatly. Consequently, robust predictions of the effects of climate change on larval supply are difficult, but investigators have emphasized the need to intensively manage the best-connected reefs as sources of larvae for recovery.

OCEAN ACIDIFICATION

As described in the previous chapter, ocean acidification has decreased seawater pH, and models predict a further reduction of 0.3 to 0.5 pH units over the next 100 years. The impacts of ocean acidification will be particularly severe for shelled organisms; tropical and coldwater corals will require more energy to calcify under these more acidic conditions. The impacts on fish are, however, just being explored, but some early experimental results indicate that higher water temperatures reduce aerobic performances of coral reef fishes and increase levels of dissolved CO_2. Considerable differences in tolerances to these conditions exist between species and these differences are expected to cause shifts in coral fish communities toward greater dominance of species tolerant of acidified seawater (Nilsson et al. 2009, Munday et al. 2009b). Studies on larval development, homing, and predator avoidance have shown that these are all influenced by seawater acidification (Munday et al. 2009a,c). In the case of larval development, the effect of acidification is to increase larval weight, which may improve survival, but acidified seawater also disrupts the ability of larvae to smell predators and find home sites (Dixson et al. 2010). Further experimentation and study is expected to uncover other life-history characteristics and species-specific responses to acidification that should illuminate the full consequences for coral reef fishes.

CHANGING PREDATOR-PREY RELATIONSHIPS

The loss of reef complexity and hiding places should have the greatest effect on small-bodied species and solitary fish (as opposed to schooling species) that become easy prey in the simplified environment. This will create bottlenecks for the recruitment of juvenile fish to reefs and the fishery, and this is expected to lead to declines in fisheries because of the low survival of juveniles and poor recruitment into the larger and fishable sizes (Graham et al. 2007). Eventually, predators may decline as their prey diminish, leading to lost productivity of the fishery (Pistorius and Taylor 2009).

ALGAL OVERGROWTH AND THE ROLE OF HERBIVORES

Turf or other forms of algae can rapidly colonize the space created by dead corals following bleaching events. Algal colonization and dominance can, however, be compensated by grazing herbivores that principally include a suite of fish species and grazing sea urchins (McClanahan et al. 2001). By intensively grazing on the early succession algae, a high abundance of herbivores can prevent the proliferation of algae after corals die, allow the settlement of coral recruits, and in some cases even reverse the dominance of algae (Bellwood et al. 2007, Hughes et al. 2007). Colonizing algae can, however, swamp herbivores when numbers are low or when corals die over large areas (Williams et al. 2001).

Herbivorous fishes, such as many surgeon and parrotfishes, prefer turf algae and may increase in numbers or grow faster when dead corals are replaced by fast-growing turf algae (Carpenter 1990, McClanahan et al. 2002, Cheal et al. 2008). In contrast, relatively few species feed on late-succession erect fleshy algae (McClanahan et al. 2001, Bellwood et al. 2007), and once erect algae have become established, they can be very persistent and even inhibit those herbivorous fishes that prefer the more palatable turf algae (McClanahan et al. 2001, McClanahan et al. 2002). The species that reduce the late-successional algae are more likely to be browsing fishes such as rabbitfish (Siganidae), rudderfishes (Kyphosidae), some browsing parrotfish (Calotomus and Leptoscarus), and even some species that are not known to primarily eat algae, such as batfish (Ephippidae; Bellwood et al. 2007).

Experimental manipulations of herbivores have repeatedly confirmed their role in controlling the abundance of erect algae and influencing the recruitment of corals and competition with erect algae (Bellwood et al. 2004, Burkepile and Hay 2006, Mumby et al. 2006). One study in the Great Barrier Reef, for example, studied coral recovery after a bleaching event by using large cages and control reefs to experimentally exclude herbivorous fish (Hughes et al. 2007). This study found that corals increased from 10% to 30% where there were no herbivores, but up to 80% where herbivores were present. A global compilation of coral cover data found that the oldest marine protected areas were more likely to recover after disturbances than were newly established MPAs, which the authors attribute to higher herbivore abundance in the older MPAs (Selig and Bruno 2010). Yet coral cover declined before recovering for those MPAs that were less than 4 and 14 years old in the Indo-Pacific and Caribbean, respectively. Again, this indicates a lag in management effects and a necessary recovery period for the fish community before corals benefited from high grazing (McClanahan et al. 2007b).

TIME LAGS

Time lags between the climate disturbance and its effects on fishes are sometimes masked by flexibility in both their diet and habitat requirements. First, as mentioned earlier, facultative species can switch to non-coral diets, but the new diet may be less healthy and eventually reduce the reproduction potential of these individuals (McIlwain and Jones 1997, Pratchett et al. 2006). Second, while these adult fishes are able to migrate away from disturbed habitats, their offspring require coral or complex reef structure as habitat for survival. Without it, these juvenile fishes suffer poor health, high mortality, and poor replacement of adults (Feary 2007). Third, species that are weakly or not dependent on coral often rely on prey that is dependent on coral. This can create a delayed decline in these predatory fish after a disturbance, as was suggested for a notable decline in predators 5 years after the high coral mortality in 1998 in Aldabra atoll, Seychelles (Pistorius and Taylor 2009). Finally, there are frequently 3- to 10-year time lags between coral mortality and the loss or collapse of reefs' topographic complexity (Graham et al. 2007). A few field studies have attributed declines in the abundance and diversity of coral reef fishes to a delay between coral death and reef structural collapse (Graham et al. 2006, Graham et al. 2008b). Time lags and the complexity of species and habitat interactions make it challenging to fully understand all of the possible effects and how they interact.

Herbivores are hypothesized to increase in abundance following coral decline due to a greater availability of algal resources. Empirical studies, however, have reported high variability, and since most have been conducted shortly after disturbances, they were unable to reach conclusions about the long-term changes (Wilson et al. 2006). A study of fisheries landings in the Seychelles indicated that a loss of small herbivores was occurring in the fishery after coral mortality and suggested that the loss of refuge from predators was impairing the recruitment of adults to the fishery (Graham 2007). This slow decline in herbivore numbers and the lost ability to check algal growth is expected to slowly degrade the reef ecosystem. Coral loss leads to an increase in a variety of algae and some other benthic groups like sponges and soft corals and not just an increase in unpalatable erect algae (Bruno et al. 2009).

Recovery

Loss of coral is likely to be bad for coral reef fish. If bleaching is infrequent, affects a small area, and does not destroy the topographic complexity of coral reefs, then even highly susceptible fish species may be able to persist and recover. For example, the optimal level of coral cover for fish diversity may be around 25% of the benthic cover (Holbrook et al. 2008, Wilson et al. 2009). Declines in fish numbers and diversity may not occur until coral cover is less than 10% of the substratum. Nevertheless, loss of coral cover can result in changes to the fish community, toward more herbivores and less corallivores. Recovery may depend on the frequency and intensity of disturbance and the recovery of highly sensitive branching coral species that provide habitat complexity (Cheal et al. 2008, Emslie et al. 2008). Some studies show that coral cover may take as little as 5 years to return to pre-disturbance levels

(Halford et al. 2004, Gardner et al. 2005). Nevertheless, many examples show that recovery is very slow, it is not occurring on the decadal scale that it has been measured on (Gardner et al. 2003, Baker et al. 2008), or the coral species composition is changed greatly after the disturbance (McClanahan 2008).

Studies of Kenyan fisheries closures, where fishing has been excluded for up to 40 years, indicate that the diversity of fish (i.e., numbers of fish species) recovers relatively quickly, on the order of 10 years, while total fish biomass takes closer to 20–25 years (McClanahan et al. 2007b). Some fish, such as surgeonfish, triggerfish, and rabbitfish, were even slower to recover and did not show signs of full recovery in even the oldest closures. Consequently, where fishing and coral bleaching interact, recovery could be protracted for species that are sensitive to both coral mortality and fishing.

A Regional Case Study

Coral mortality after 1998 was highly variable across the Indian Ocean region, and a large-scale study of eight countries in the WIO seven years after the 1998 event also concluded that mortality was variable for coral reef fish (Graham et al. 2008b). Numbers of fish species at these study sites were weakly related to the loss of coral, with the largest losses of fish species in the Seychelles. The largest declines were among obligate corallivores, followed by planktivores, and small-bodied individuals of many species, including herbivores (Fig. 5.2). The decline in small-bodied planktivores was most likely due to their vulnerability to predators once the refuge created by the structural complexity of corals was lost (Graham et al. 2006). The observed reduction of small-bodied herbivorous species that perform important functional roles on coral reefs is a major concern because the replacement of adults will be very important for maintaining reef ecology and fisheries.

Larger herbivores and mixed diet feeders were not affected across this 7-year period. The mixed-diet group includes important fishery species such as emperors, snappers, goatfish, and wrasses. Many of these species are habitat generalists that forage and recruit to non-coral reef habitats such as mangroves, sand, and seagrass. Species in these groups also tend to forage over large spatial scales, indicating weak reliance on specific habitat locations or types. Consequently, these groups are most likely to be sustained in the long term, although high variation is expected at the species level, which will lead to changes in the fish community (Wilson et al. 2006, Cheal et al. 2008). Fortunately for our WIO study region, recovery of corals has been faster than in other ocean basins (Baker et al. 2008). In the Caribbean, many reefs are showing no or slow overall recovery, although recovery was shown to be faster in areas with restricted fishing (Selig and Bruno 2010).

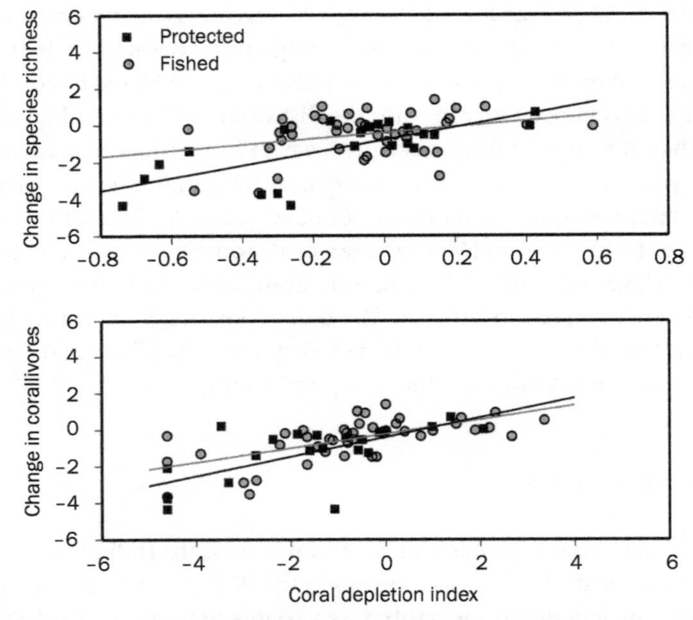

FIGURE 5.2 Relationship between coral depletion 7 years after the 1998 event and changes in numbers of fish species and corallivores, which were the two strongest studied variables in this multinational study.

Source: Graham et al. (2008b).

Impacts of Climate Change on Reef Fisheries

Worldwide, the goods and services that coral reefs provide are estimated to be worth almost U.S. $30 billion per year, and a significant amount of this value is derived from reef fisheries (Cesar et al. 2003). Fisheries production could be affected by climate change through the numerous mechanisms described in this chapter. Despite the potential enormous losses, only a few studies have empirically examined the effects of severe and large-scale coral mortality on the value of fisheries (Pratchett et al. 2008). Additionally, fishing often has a very strong effect on coral reef ecology, and this makes it difficult to definitively detect the effects of climate change and coral mortality when both disturbances occur simultaneously (Wilson et al. 2008).

The most informative studies include long-term trends on fishing effort, catch composition, and habitat characteristics because these can control for changes in fishing effort and management. For example, studies in Kenya found declines in catch after coral bleaching events, but either the declining patterns were present before the bleaching or they were attributed to changes in fishing effort (Westmacott et al. 2000a,b, McClanahan et al. 2002). Likewise, in the Seychelles' fishery, low abundance and yield of herbivorous rabbitfish was attributed to a pre-bleaching trend associated with fishing effort (Grandcourt and Cesar 2003). A very lightly used fishery that targets species strongly associated with coral reefs did, however, find a decline in catch of some predatory species 5 years after the 1998 coral mortality event (Pistorius and Taylor 2009). Another long-term study across this period found that catches improved after restrictions on small-meshed nets were in place (McClanahan 2010). Consequently, catch responses may be small when fisheries target herbivorous or generalist species; time lags are expected for many species, but improved management can potentially offset catch losses.

Many tropical fisheries' target fishes are not dependent on coral or the hard-bottom benthos. For example, the overwhelming majority of fish caught in Kenya and Papua New Guinea artisanal fisheries were associated with the reef structure but not explicitly reliant on live corals (Cinner et al. 2009a). A significant portion of landings in tropical fisheries is caught in the proximity but is not fully dependent on coral reef habitat. These include, for example, jacks (Carangidae) and other pelagic fishes (Engraulidae, Clupeidae, Scombridae) that can temporarily congregate on reefs. Moreover, climate change effects are not simply restricted to coral mortality but can also influence the frequency and strength of cyclones, upwelling, rainfall, runoff, and sedimentation (McClanahan 2002b). Consequently, effects of climate-induced coral bleaching on coral reef fisheries are likely to be difficult to detect, as they are highly protracted or part of multiple disturbances, and therefore potentially masked by a wide range of other factors.

The most apparent effects may be among niche fisheries such as the aquarium trade that target mainly coral-dependent fishes. The international marine ornamental fish trade is currently worth U.S. $90–$300 million per year and mostly targets small coral reef fishes, including butterflyfish, cardinalfish, damselfish, and filefishes (Sadovy and Vincent 2002), of which many are highly susceptible to climate-induced coral bleaching (Graham et al. 2011). The recent disappearance of several coral-dependent fishes from aquarium catches, including the beaked leatherjacket (*Oxymonocanthus longirostris*), has been directly attributed to declines in stocks following the 1998 mass-bleaching event (Dulvy et al. 2003). Climate-induced coral bleaching may eventually affect overall catches of aquarium fishes and possibly the total value of the trade.

Conclusions

Fish are sensitive to fishing and disturbances to their environment and habitat, but the response can be complex and difficult to predict in terms of species-specific responses due to habitat, food, prey and predator interactions, size specificity, and time lags. These are common problems for understanding ecological responses to all types of disturbances and even more so for rapid climate change, which will influence not just seawater temperature but also a whole suite of factors from water chemistry to oceanic circulation. Regardless, the recently accumulating evidence is indicating a considerable number of changes that will ultimately influence fisheries catches in terms of species composition and yields. The effects of reducing fish biomass on the ecology of the coral reefs are evident throughout the region as shown above. But the following chapters will make it clear that social systems are major drivers of ecological change and that management can drive ecological processes and resilience as much as or more than climate disturbances. The next two chapters further examine socioeconomic conditions in the region and the ways in which people cause and respond to ecosystem change.

6

Vulnerability of Coastal Communities

The key impacts of climate change in the western Indian Ocean (WIO) area are likely to be (1) widespread degradation in coral reef ecosystems from coral bleaching and ocean acidification, (2) changes in productivity of rain-fed crops and foraging livestock, (3) reduced water availability, (4) increasing sea-level rise and cyclone paths, and (5) changing distribution and increased outbreaks of diseases that will affect humans, livestock, and crops (Thornton et al. 2008). These climate change disturbances are not only likely to vary from place to place, as described in previous chapters, but will also vary for different people within society, depending on their level of vulnerability.

Vulnerability is the level of susceptibility to harm from events such as coral bleaching, cyclones, and sea-level rise (Gallopin 2006). It is a measure of human well-being and is fundamentally linked to people's socioeconomic conditions, including such factors as where they reside, their dependence on natural resources, and their access to resources for coping with disturbances (Adger 1999, 2006). In the technical literature, vulnerability is often described as having three inter-related components: exposure, sensitivity, and adaptive capacity (Fig. 6.1). One common way to characterize people's vulnerability is to examine and compare key socioeconomic indicators that have been evaluated by theory and empirical research (Adger and Vincent 2005, Adger 2006, Allison et al. 2009).

As an example at the global scale, Allison and colleagues (2009) developed a national-level measure of vulnerability to the impacts of climate change on fisheries. This measure combined national-level variables such as fish catch production, fisheries exports, the number of people working in aquaculture and capture fisheries, per capita gross domestic product (GDP), and dependence on fish as a primary source of protein. Of the 15 countries with the highest vulnerability of fisheries to climate change, 13 or 14 were in Africa, depending on the severity of the climate change scenario used. Many countries were in western Africa, but Mozambique was highly vulnerable under both emissions scenarios. This study is useful for the big-picture view of how relatively vulnerable the wider economy of nations can be to the effects of climate change on fisheries. The trade-off inherent in taking this big picture view is that national-level evaluations and indicators often fail to account for the considerable heterogeneity that occurs at community and household scales, where some key adaptation decisions are made.

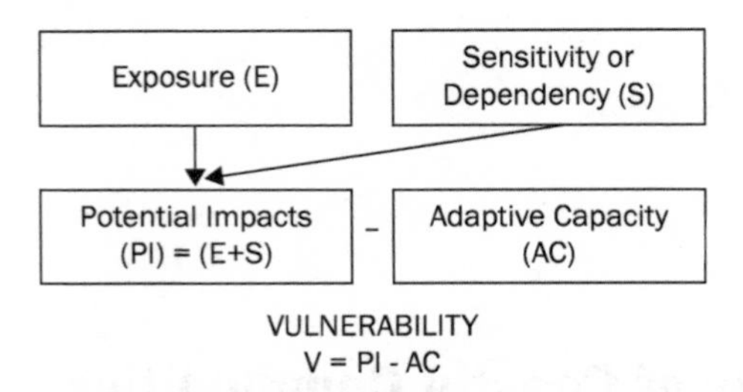

FIGURE 6.1 Vulnerability is comprised of exposure and sensitivity (which captures the potential impacts) and adaptive capacity (which captures people's ability to cope with or adapt to change).

Source: Adapted from Allison and colleagues (2005).

This chapter builds on existing national-scale studies of vulnerability to climate change (including those developed by, for example, Adger and Vincent 2005, Brooks et al. 2005, and Allison et al. 2009) and explores both local and national-scale indicators of vulnerability to climate change. In Chapter 8, a local scale index of two key components of vulnerability, exposure and adaptive capacity, is developed and quantitatively compared throughout the region.

Exposure

Exposure is the degree to which a system is stressed by the environment. This can be characterized by the magnitude, frequency, duration, and spatial extent of a climatic event, such as coral bleaching or a cyclone. Exposure may vary based on factors such as oceanographic conditions, prevailing winds, and latitude. These variable factors result in some areas having a higher likelihood of being affected by events such as cyclones or coral bleaching. Key aspects of exposure to physical climate change events such as cyclones, coral bleaching, and sea-level rise were described in previous chapters. These chapters demonstrated the high spatial variability of different physical elements of the environment. In particular, rapid climate change is likely to produce more temperature-induced coral bleaching in the northern WIO and more tropical storms and cyclones in the southern WIO in the coming decades. Rainfall patterns at the coast will vary, with the southern African region drying and the equatorial region having stronger short and weaker long rains.

For certain climate change disturbances, the location or settlement pattern of communities will influence their exposure. An obvious example is that communities located away from the immediate coast or at elevation will be only indirectly affected by sea-level rise or events such as storm surges arising from cyclones. The adaptations that societies can undertake to minimize exposure are limited. These adaptations primarily rely on engineering solutions, such as erecting levees or sea walls to mitigate coastal disasters; coastal planning, such as building away from danger zones; and disaster preparation, such as early warning systems for cyclones.

Sensitivity

Sensitivity is the degree to which a disturbance or stress, such as a coral bleaching event or cyclone, modifies or affects a system, and there is both biological and social sensitivity. An example of biological sensitivity is a species going extinct when the climatic conditions needed for successful reproduction are exceeded. Social sensitivity, which is the main focus of this chapter, may be affected by conditions such as local-level dependence on natural resources. Low dependence on fishing or other marine resources may mean, for example, that climatic events such as coral bleaching have a minimal economic impact on coastal communities.

The main sectors that are likely to be directly affected by climate change are agriculture, tourism, and fisheries. The contribution of these sectors to WIO national economies varies greatly (Fig 6.2). Agriculture, in particular, comprises a significant proportion (>20%) of the national GDP of Kenya, Madagascar, Mozambique, Comoros, and Somalia. In every country except Kenya, agriculture contributes more to the numbers employed in the workforce than the monetary contribution toward the national GDP (Fig. 6.2). This suggests that agriculture represents an important livelihood in the region, and with more than 60% of the workforce in Tanzania, Madagascar, Mozambique, Comoros, and Somalia employed in agriculture, these countries will be highly sensitive to changes in food production.

In the Seychelles, Mauritius, and the Maldives, tourism is a large contributor to both workforce employment and GDP. These areas are highly sensitive to environmental changes, such as degradation of the environment by coral bleaching, which could make these countries less attractive as ecotourism destinations compared to locations without bleached and dead corals. It is estimated that the total international tourist arrivals to sub-Saharan Africa, including the Indian Ocean, will increase from about 27 million in 2000, to 47 million in 2010, and to 77 million by the year 2020, with southern and eastern Africa and the Indian Ocean islands experiencing the fastest growth (WTO 2001). These projections greatly depend on the global economies and costs of long-distance air transportation but do suggest an increasing dependency on tourism in this area.

At a national level, the fisheries sector does not appear to contribute greatly to the GDP or workforce employment anywhere but the Seychelles and to a lesser extent the Maldives and Comoros. Nevertheless, since many

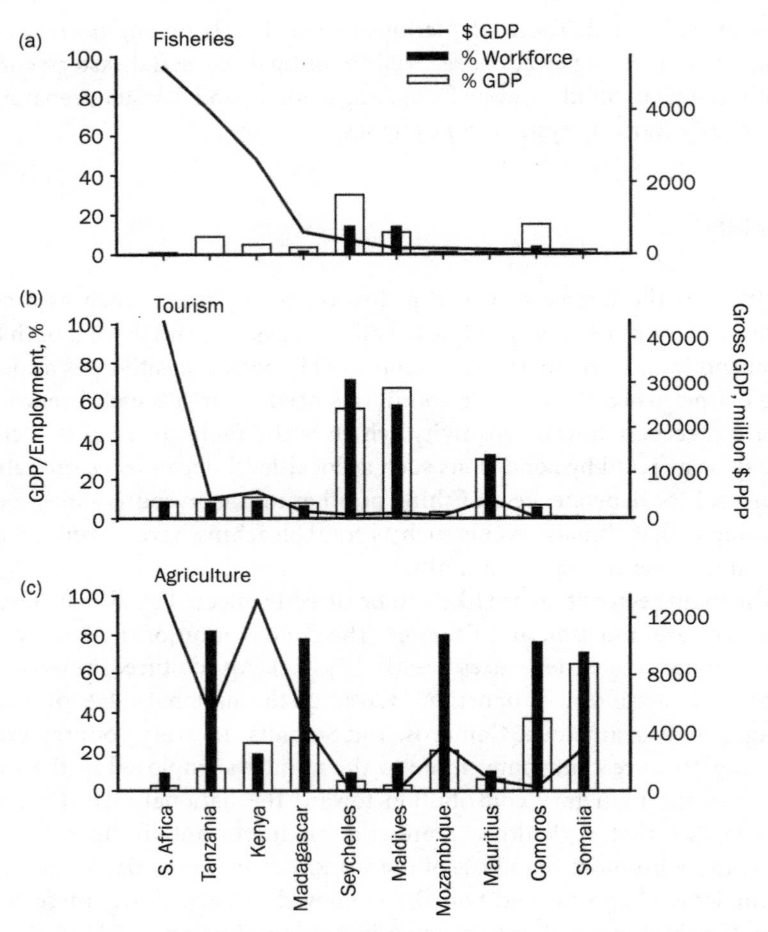

FIGURE 6.2 Percentage of people employed, total value, and proportion of gross domestic product (GDP) of (a) fisheries, (b) tourism, and (c) agriculture to countries in the WIO. PPP: purchasing power parity or corrected cost of products in different countries.

of the countries are large with sizable inland populations and small-scale coastal catches not well quantified, national-level statistics do not properly reflect the sensitivity of coastal communities to changes in the fishery. An example of the mismatch between country-level data and community-level realities is exemplified by our study of more than 1,600 households from 29 communities throughout coastal Kenya, Tanzania, Madagascar, Seychelles, and Mauritius (McClanahan et al. 2008b, Cinner et al. 2009b, Cinner et al. 2009c). Despite national statistics that suggest fisheries are not particularly important in Kenya, Tanzania, and Madagascar, the level of dependence on fishing as either a primary or secondary occupation was extremely high in a number of coastal communities in these countries (Fig. 6.3) Alternatively,

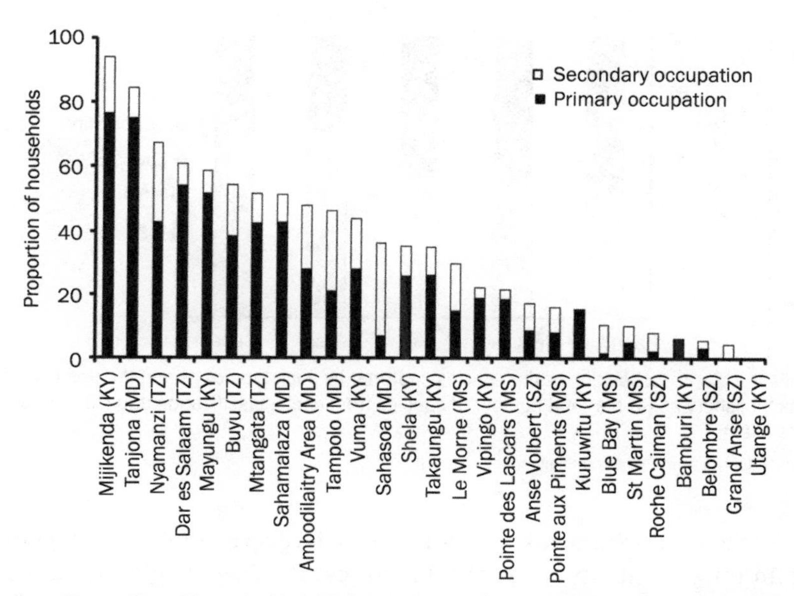

FIGURE 6.3 Proportion of households in the western Indian Ocean that are involved in fishing as a primary or secondary occupation. MD -Madagascar, KY - Kenya, TZ - Tanzania, MS - Mauritius, SZ - Seychelles.

Source: Adapted from Cinner and Bodin (2010).

when compared to the high importance of fishing in the Seychelles' national-level statistics, Seychellois communities had very low levels of dependence on fishing as an occupation.

Other factors that may influence sensitivity to climate change include the use of specific technologies. For example, there are differences in the catch of fish sensitive to coral bleaching among the different fishing gears used in Kenya (Fig. 6.4; Cinner et al. 2009a). Fish captured in the multispecies fisheries have differing levels of association with the corals themselves. These fish can broadly be placed into three categories: (1) some fish are highly associated with reefs and entirely dependent on coral for feeding, dwelling, and settlement; (2) some fish use the coral complex as habitat but are not entirely dependent on the reef for key life stages; and (3) some fish are not strongly associated with the reef. Using these broad categories of reef association helps identify species most affected by coral bleaching and associated mortality. Fishes with strong associations are mostly likely to be affected by coral bleaching or other mortality events. Fishes with moderate associations are likely to be affected by a collapse of the reef structure, which may occur 7 to 10 years after a mortality event (Graham et al. 2007). Importantly, fishers employing gear such as spears and traps are more likely to be negatively affected by climate change events that kill corals.

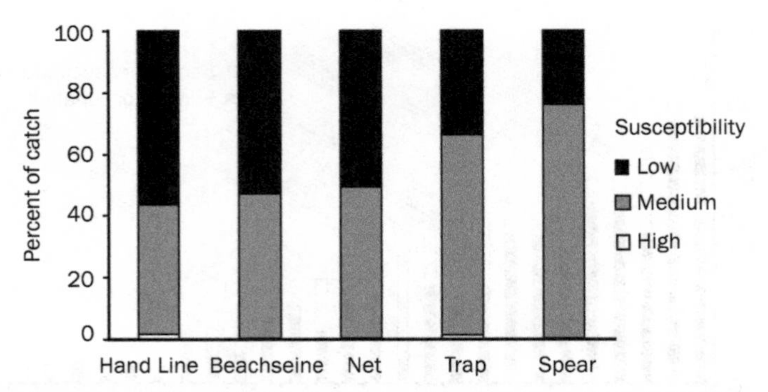

FIGURE 6.4 Coral association of the species captured by each type of gear in the Kenyan artisanal fishery as (a) the total number of species from each classification and (b) a proportion of the catch.

Source: Adapted from Cinner and colleagues (2009d).

Societies, governments, and donors can develop a number of adaptations to minimize sensitivity to certain climatic events. These might include diversifying livelihoods to reduce dependence on local natural resources (Allison and Ellis 2001) and diversifying fishing operations to minimize the use of gear that specifically targets fishes that are more likely to be affected by climate change.

Adaptive Capacity

Adaptive capacity refers to the conditions that enable people to (1) anticipate and respond to change, (2) minimize and recover from the consequences of change, and (3) take advantage of new opportunities (Adger and Vincent 2005, Grothmann and Patt 2005, McClanahan et al. 2008b). Theoretical and empirical research suggests that people with high adaptive capacity are less likely to suffer from climate change and are better able to take advantage of the opportunities it creates. A number of studies have investigated the socioeconomic determinants of adaptive capacity, which can broadly be grouped into four key aspects (Cinner et al. 2009b). These are (1) flexibility (Gunderson 1999, Adger 2000), (2) assets (Adger 2000), (3) learning (Carpenter et al. 2001, Lebel et al. 2006), and (4) social organization (Carpenter et al. 2001, Smit and Pilifova 2003, Lebel et al. 2006). These dimensions of adaptive capacity are explored next.

FLEXIBILITY

Flexibility of individuals and institutions is a critical component of adaptive capacity. Understanding where sources of flexibility already exist and where it can be increased is vital to effectively build and manage the resilience of social-ecological systems. In many coastal communities, the flexibility to switch between livelihood strategies is critical if resource users are to cope with the high uncertainties and seasonal variability associated with fishing, farming, and other coastal livelihoods (Allison and Ellis 2001, Berkes and Seixas 2006, Hoorweg et al. 2008). Flexibility in livelihood strategies allows resource users to switch to alternative sectors, occupations, or fishing gear to "smooth" the effects of temporal variations in the accessibility of resources (Allison and Ellis 2001).

One indicator of livelihood flexibility is whether households are engaged in multiple occupations. There are two broad perspectives about the causes and consequences of livelihood diversification strategies. One view is that a progression from low to high standards of living involves a transition from diversification to specialization (Bernstein et al. 1992, Davies 1996). In this context, livelihood diversity generally reflects a low standard of living because it is necessary to fill a gap between consumption and food production. An alternate view suggests that diversification is a deliberate strategy adopted by proactive households, based on the principle of the "portfolio" or the spreading of risk (Mortimore 1989, Stark 1991, Allison and Ellis 2001). This perspective suggests that adopting a diverse livelihood portfolio can improve resilience to seasonal, annual, acute, and chronic fluctuations and shocks. These two perspectives are not necessarily mutually exclusive, but they do differ considerably in regard to the merits they place on livelihood diversity.

In coastal communities throughout the WIO, households generally have several livelihood activities to draw on, ranging from an average of 1.5 (±0.1 standard error) person-jobs per household in Pointe de Lascars and St. Martin (in Mauritius) to a high of 6.7 (±0.4 SE) in Tampolo, Madagascar (Fig. 6.5). These numbers are the total number of jobs conducted by all members of the household. For example, one person who both goes fishing and engages in farming would contribute two person-jobs to the statistics for the household. This trend does not change when the size of a household is controlled for. In this example, we have used broad economic sectors (fishing, salaried employment, farming, cash crops, etc.) as our measure of work (see Cinner and Bodin 2010 for more details).

There are considerable country-level differences in the diversity of livelihood portfolios. For example, households in Madagascar stand out for their high number of jobs (5.0 person-jobs per household) compared to countries such as Mauritius (1.7 person-jobs per household). In Madagascar, this high level of "occupational multiplicity" may reflect coping or risk-reduction mechanisms to deal with high levels of economic, political, and environmental variability. For example, since 2000, Madagascar has experienced a combination of disturbances, including two presidential crises (in 2002 and 2009), a 10-fold price drop in vanilla in 2005,- which is an economically important

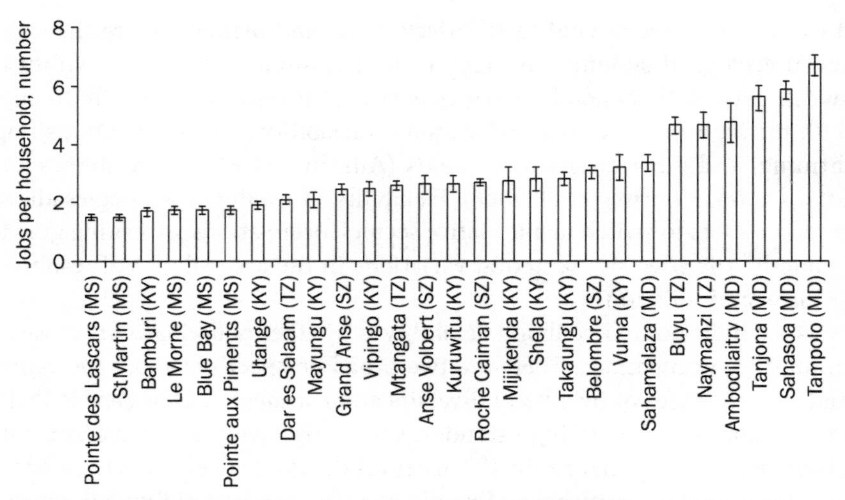

FIGURE 6.5 Mean number of occupations per household (+ standard error) in 27 coastal communities in the western Indian Ocean.

Source: Adapted from Cinner and Bodin (2010).

cash crop (ITC 2007), and six major cyclones in 2006–2007. These political events caused multiple suspensions of international flights and the cyclones left nearly 500,000 people requiring humanitarian assistance (Duffy 2006, Reuters 2007). Frequent disturbances like these may require most households to have a high risk-spreading strategy to survive.

There is also considerable variation in livelihood portfolios within countries. Households in Sahamalaza, Madagascar, on average engage in almost one less occupation than households in the other parts of Madagascar. This suggests that livelihood portfolios in Sahamalaza are either less flexible or the occupations they have are more stable or profitable. Urban-rural trends in occupational multiplicity explain some of this variation within countries. Rural sites in Kenya, such as Vuma and Takaungu, and Tanzania, such as Buyu, had a considerably higher number of occupations per household than the peri-urban areas of Mombasa, Kenya (Bamburi and Utange), and Dar es Salaam, Tanzania (Fig. 6.5).

A range of factors, including national and international donor macroeconomic policies such as subsidies, can affect the flexibility of livelihood portfolios. The removal of agricultural subsidies in Tanzania diversified nonfarm incomes (Bryceson 2002). Supplemental income projects and alternative fishing strategies may be a key focus for increasing flexibility and reducing vulnerability to climate change impacts in coastal regions (Allison and Ellis 2001). Supplemental income projects can result in disappointing and sometimes even perverse outcomes when the social, economic, and cultural context of fishers are not adequately considered (Pollnac et al. 2001, Sievanen et al. 2005). An example of a perverse outcome can be that supplemental activities may allow people to subsidize a culturally important activity such as fishing, even when it is not economically viable.

Institutional flexibility also provides the ability to respond to change and is key for developing adaptive experimental behavior (Tompkins and Adger 2004, Berkes and Seixas 2006). In parts of the WIO, such as Kenya, Tanzania, and Madagascar, communities are being increasingly empowered to manage their local coastal resources. Community-based management arrangements, known locally as beach management units (BMUs) in Kenya and Tanzania and the Gestion Locale Sécurisée (GELOSE) in Madagascar, are becoming pervasive and often provide more management flexibility than national-level governance (Cinner et al. 2009d).

Local resource use restrictions frequently occur as resource-habitat taboos (Ruud 1960, McClanahan et al. 1997, Cinner et al. 2009b). Studies in Papua New Guinea and Indonesia show that in some places, these customary management practices can be highly adaptive and conserve marine resources (Cinner et al. 2006). In contrast, customary taboos on resource use in Kenya (McClanahan et al. 1997) and Madagascar (Elmqvist 2004, Bodin et al. 2006, Cinner 2007) appear to be relatively inflexible institutions that focus on spiritual connections to ancestors rather than adaptively manipulating resources (Cinner and Aswani 2007). In these cases, local sociocultural institutions may actually inhibit flexibility.

SOCIAL ORGANIZATION

A society's ability to adapt is determined in part by its ability to organize and act collectively (Adger 2003). Organization and collective action enable people to respond to disturbances by drawing on resources outside of their households. This aspect of vulnerability is influenced both by the effectiveness of institutions and social networks and by demographic trends, such as population growth and migration as well as other factors (Smit and Pilifova 2003, Thornton et al. 2008).

Governments and nongovernmental organizations (NGOs) often provide the basis for social organization and deliver services that can play a critical role in adapting or responding to a disturbance. Ineffective governments, political conflicts, corruption, and a lack of accountability can, however, hamper organization and delivery of services during periods of disturbance and climate stress (Barnett and Adger 2007). Throughout parts of the WIO, these types of governance issues result in high levels of vulnerability (Table 6.1; Brooks et al. 2005, Kaufmann et al. 2005, Thornton et al. 2008). Mauritius, Seychelles, and South Africa stand out as ranking high (50th to 70th percentile) in the world for four important indicators of good governance (Table 6.1). In contrast, Somalia, Comoros, and Kenya rank lowest in the world in regard to key aspects of governance, such as government effectiveness, corruption, and

TABLE 6.1

Ranks of western Indian Ocean countries relative to the rest of the world in key governance indicators

Country	Government effectiveness	Control of corruption	Political stability	Voice and accountability
	Percentile rank (0–100)	Percentile rank (0–100)	Percentile rank (0–100)	Percentile rank (0–100)
Comoros	0.9	29	30.3	34.6
Kenya	30.3	15.5	15.9	46.2
Madagascar	46.9	56.5	40.4	47.6
Mauritius	72	69.6	71.6	72.6
Mozambique	40.3	35.3	57.2	47.1
Seychelles	55.9	59.9	83.7	46.6
Somalia	0	0	0	3.4
South Africa	74.9	67.1	51	68.8
Tanzania	39.3	43	39.9	43.8

The highest possible rank is 100.
Source: World Bank (2010).

stability (Table 6.1). The Intergovernmental Panel on Climate Change (IPCC) notes that many of the African institutional and legal frameworks are insufficient for dealing with disaster risks (Boko et al. 2007). Importantly, in Africa and throughout the world, it is often the most vulnerable groups that are excluded from formal decision-making processes concerning management of climate-related risks (Adger 2003).

Civil society, which can include institutions such as NGOs; trade unions; churches; and student, business, or professional associations (Makumbe 1998), can be crucial in how societies organize and adapt (Adger 2003). Where governments are weak, civil society plays a primary role in adaptation to and recovery from disturbances (Adger 2003, Thomas et al. 2005). For many decades, civil society has played a key role in the social and political fabric in much of Africa (Makumbe 1998). Recent political and macroeconomic shifts, such as decentralization and structural adjustments, respectively, have led to an increased role for local civil society (Hara and Raakjaer Nielsen 2003). In parts of the western Indian Ocean area, such as Kenya and Madagascar, community-based organizations are playing an increasingly active role in managing natural resources (Cinner et al. 2009d).

Another key part of people's adaptive capacity is their ability to act collectively (Adger 2003). Critical to this ability are the interactions and bonds within a society—a concept referred to as social capital. Adger (2003) identifies two key types of social capital that are particularly important in adaptation: (1) bonding capital, which focuses on intra-community social bonds and is particularly important in low-income groups; and (2) bridging or networking capital, which focuses on extra-community social bonds and is particularly important when dealing with the absence of strong governments. The second type of social capital is also important with dynamic and mobile communities, such as migrant fishers, and when managing common pool resources, such as coral reefs and associated fisheries.

Some aspects of social organization can impose constraints on people's ability or willingness to adapt. For example, in the Pulicit Lagoon in India, some fishers were unwilling to adapt their fishing techniques during poor harvests because adaptive fishing gear was associated with lower social castes (Coulthard 2008). The presence of certain sociocultural norms may also inhibit people's ability to engage in certain adaptations. Customary norms or taboos in much of coastal Madagascar, including sacred areas and behavioral restrictions, may restrict where people can fish, the days that people can farm or fish, the types of fishing gear they can use, and the species they can target (Table 6.2). In both of these cases, a number of potential adaptation options

TABLE 6.2

Presence of specific taboos mentioned by fishermen and key informants in four study areas of Madagascar

Description	Tanjona	Cap Masoala	Tampolo	Sahamalaza	Nosy Antafana
Sacred areas					
Sacred area		X		X	X
Food					
Guitarfish	X	X	X	X	
Turtle	X	X	X	X	
Puffer fish or eggs	X		X	X	
Dugong	X	X	X		
Red parrotfish	X	X			
Other*	X	X	X	X	
Time					
Limit work in fields**	X	X	X	X	X
Fishing on Saturday				X	
Gear					
Traps	X	X		X	
Speargun			X		X
Weir				X	
Use black line/rope				X	
Total	14	11	11	12	4

* Electric and poisonous fish, cardinal fish, juvenile fish, stingray, octopus.

** On specific days: Thursday, Tuesday, Monday, Sunday, and Wednesday.

Taboos are grouped into four broad categories: Sacred areas, food, time, and gear. Taboos may be permanent or temporal. X = taboo present.

Source: Cinner (2007).

may be unavailable as a result. Social and cultural constraints can play a large role in people's willingness or ability to adapt but may not be easily perceived by people who are unfamiliar with local traditions, customs, and social structure. Consequently, these types of constraints are often missing in indicator-based assessments of vulnerability.

Demographic issues such as population and migration can also influence the ability to organize. Ostrom (1990) notes that cooperation around common property resources is more likely among small social groups. Yet human populations in this region are set to expand from 0.9 billion in 2005 to 2.0 billion by 2050 (Thornton et al. 2008). Countries in the WIO have highly variable populations and growth rates, ranging from about 40 million (with about 2.5% growth rates) in Kenya and Tanzania to 87 thousand (with a 0.4% growth rate) in Seychelles (Table 6.3). South Africa has the highest population, of 48 million, but a low growth rate of 0.5%. For large countries with sizable inland populations, national-level statistics can be quite meaningless. In Kenya, less than 10% of the population (roughly 3 million people) lives on the coast, while in Tanzania coastal dwellers total just over 20% or 8.5 million people. Coastal population densities are highest in the small island states such as the Maldives, Mauritius, Comoros, and Seychelles, with between 191 and 1,230 people per km^2 compared to an average of just 19 to 25 per km^2 in Kenya and Tanzania and 2.0 per km^2 in Mozambique. Of course, there are urban and peri-urban areas in these latter countries where local population density is very high.

TABLE 6.3

National and coastal populations in the western Indian Ocean

Country	Population			
	Total (thousands)	Coastal (thousands)[Ω]	Coastal density (people/km^2)[Ω]	Total population growth (%)[◊]
Seychelles	87	87	191	0.4
Comoros	839	839	420	2.3
Maldives	369ℓ	369	1230	1.8
Tanzania	40,454	8495	25	2.4
Kenya	37,538	3003	19	2.6
Madagascar	19,638	10801	11	2.6
Somalia	8699	4784	12	2.8ʃ
Mauritius	1262	1262	631	0.7
Mozambique	19,792	11677	2	1.8
South Africa	48,577	18,945	51	0.5

[#] OECD (2007); [Ω] World Resources Institute (2003); [◊] Human development reports (2007); ℓ CIA World Fact.

Migration may be used as an adaptive measure in response to shocks such as conflicts or climate stress, and money sent home from migrants can provide critical resources for coping after a disturbance (Boko et al. 2007). However, high levels of in-migration can affect community heterogeneity (Curran and Agardy 2002) and can make it difficult for a society to collectively organize around the use and management of common-pool resources (Ostrom 1990).

Migration is critical in structuring human populations and livelihoods in Africa. In sub-Saharan Africa, migration may include seasonal, contractual, short-term, long-term, or permanent migration (Njock and Westlund 2010). Mass migration is often related to economic, political, religious, or environmental factors that include natural disasters and environmental change. Seasonal migration is common in Kenya and Tanzania, where people move depending on the cycle of planting and harvesting of key crops (World Bank 2006). More recently, health issues, including the spread of diseases, such as the human immunodeficiency virus (HIV) and an accompanying rise in acquired immune deficiency syndrome (AIDS) and other associated diseases, are more prevalent among these migrating people (Torell et al. 2006).

International migrants are currently a relatively small proportion of the WIO population (Table 6.4). Within-country or domestic migrants made up 16% to 100% of populations in the 29 WIO coastal communities studied by McClanahan and colleagues (2008d). The proportion of migrants in these communities is highly variable within and between countries. Many of the

TABLE 6.4

International migrants as a percentage of the population in the western Indian Ocean, the proportion of those that are refugees, and the average their annual growth rate between 2000 and 2005

Country	International migrants as a percentage of the population	Refugees as a percentage of international migrants	Growth rate of the migrant stock (percentage)
Comoros	8.4	0	2
Kenya	1	69.9	1
Madagascar	0.3	0	0.5
Maldives	1	0	1.4
Mauritius	1.7	0	5.8
Mozambique	2.1	0.2	2
Réunion	18.1	0	5.9
Seychelles	6.1	0	1.7
Somalia	3.4	0.1	51.1
South Africa	2.3	2.6	1.6
Tanzania	2.1	73.1	–2.4

lowest migration sites were in Kenya and Madagascar, but in Tanjona and Tampolo, Madagascar, and Mijikenda, Kenya, populations were 67% to 83% migrants. International remittances from those who have emigrated can be a major factor in the flow of cash in the WIO. Estimates of worldwide remittances to developing countries range between $100 and $150 billion (Maimbo 2006). Somalia's stateless, conflict-ridden economy persists in large part because 40% of household incomes come from remittances, or money sent by family members who have migrated (Maimbo 2006).

LEARNING

Adaptation also rests on the ability of individuals, institutions, and societies to recognize change, attribute this change to their causal factors, and assess potential response strategies (Smit and Pilifova 2003, Brooks et al. 2005, Fazey et al. 2007). Recognizing change and understanding that people cause both ecosystem degradation and improvement is critical for recognizing opportunities for experimentation with resources (Cinner et al. 2009b). If people do not perceive connections between human activities and the condition of the resources they depend on, they are not likely to support management initiatives that restrict or manage resource use. In some cases, the capacity to learn can be constrained by cultural or religious views that attribute all change to supernatural agents or natural processes too variable to control. These views can be promoted by low levels of education, poverty, an unpredictable ecological and economic environment, corrupt governments, psychological defense mechanisms, and poor integration of local and scientific knowledge (Norris and Inglehart 2004, Diamond 2005, Fazey et al. 2007).

African communities have traditionally coped with changing environments and have developed an associated range of knowledge systems and practices (Boko et al. 2007). Throughout the western Indian Ocean region, there are local knowledge and learning systems including traditional ecological knowledge, folk taxonomies, and indigenous mental models of human-environment interactions (Boko et al. 2007). Recognizing new types of change involves being open to new knowledge and being able to integrate it with local knowledge (Tengo and Hammer 2003, Aswani and Hamilton 2004, Berkes and Seixas 2006).

Coral bleaching events experienced since the 1980s are largely unprecedented, based on environmental records over the past 350 years and therefore this disturbance represents a new environmental phenomenon (Zinke et al. 2008). Ecological monitoring and measurements of coral bleaching occur at many of the region's marine conservation areas, but feedback about the condition of corals and fisheries is rarely reaching communities with recommendations for appropriate management changes. Even where scientific information is shared with communities, low levels of literacy and education create challenges for the residents in understanding and synthesizing this scientific information with local knowledge (Table 6.5). Thus the potential for communities to adapt behavior and regulations based on this new information is not being fully realized.

ASSETS

Assets help people adapt to environmental changes by providing resources to draw on during times of change. Assets can also help purchase technologies to cope with change, provide monitoring and early warning technologies, and create infrastructure and housing that may withstand storms and other forms of environmental change. A key aspect of adaptive capacity relates to people's social disadvantage or lack of power to access assets, which can prevent them from mobilizing the necessary resources to cope with change (Adger 1999, Adger and Kelly 1999, Barnett 2001).

Of the 1.3 billion poor people in the world, nearly 300 million live in sub-Saharan Africa (Thornton et al. 2008). Poverty is endemic in parts of the WIO. In much of the region, extreme poverty characterizes much of the coastline and people have little access to personal or government resources for coping with change. The case is considerably different for the island nations

TABLE 6.5

Illiteracy rates and primary school attendance in the western Indian Ocean

Country	Education	
	Illiteracy rates %	Primary attendance %
Seychelles	8	99*
Mauritius	16	95*
Maldives	4	79*
Comoros	57#	31
Madagascar	29	76
Kenya	26	79
Tanzania	31	73
Mozambique	61	60
Somalia	–	20
South Africa	12	82

Illiteracy rates based on adults aged 15 and over unable to read; primary school attendance is the percentage of children of primary school age attending primary school; * OECD (2010); *Primary school enrollment, which is expected to be higher than actual attendance.

Source: United Nations (2005) except where indicated.

of Seychelles, Reunion, and Mauritius and this disparity among countries provides a broad overview of this component of adaptive capacity.

At the national level, the human development index (HDI) measures a country's achievement in terms of health, knowledge, and standard of living and thus is a good indicator of human access to key assets. Countries vary considerably within the WIO in terms of their human development. Using the HDI, the Seychelles and Mauritius have been classified as having high human development; the Maldives, Comoros, South Africa, Madagascar, and Kenya have medium human development; and Tanzania, Mozambique, and Somalia rank low. A child in Mauritius would be expected to live to the age of 72 whereas in Mozambique a child is expected to live only to 43. The average

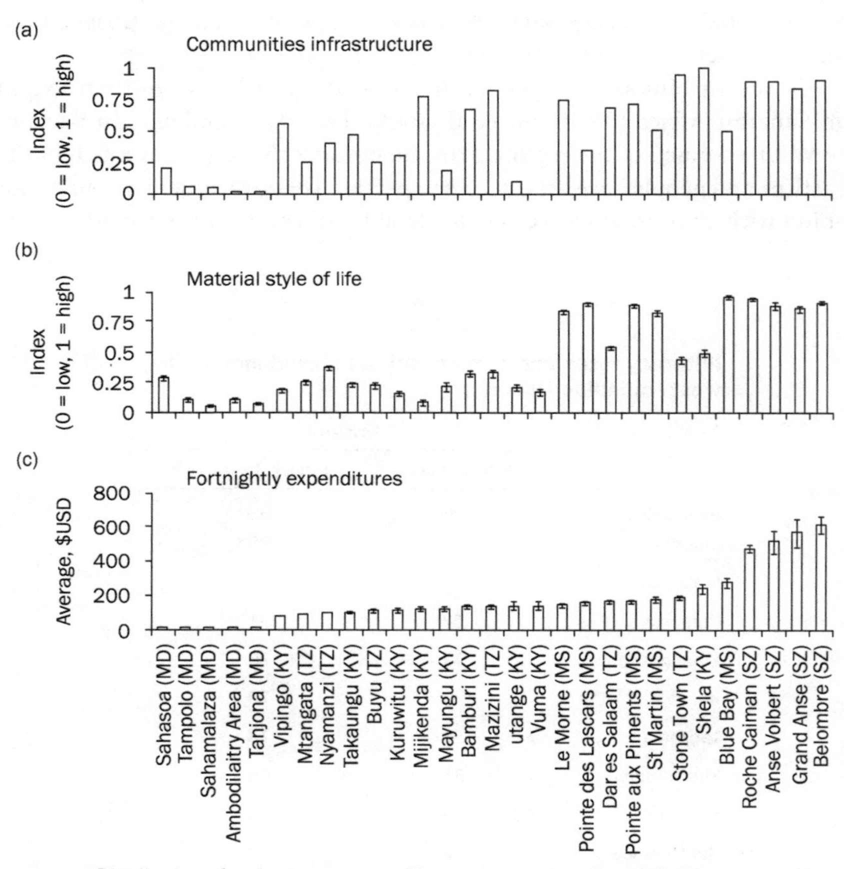

FIGURE 6.6 Distribution of various measures of human assets in selected fishing communities in the western Indian Ocean. These include (a) a relative measure of the communities' infrastructure, (b) the average household's material style of life, and (c) the household's average fortnightly expenditures.

Source: Adapted from McClanahan and colleagues (2008d).

Mauritian is expected to earn U.S.$12,700 annually, corrected for purchasing power parity (PPP), compared to only U.S.$ 1,240 (PPP) in Mozambique. In Mauritius, 84% of the population over the age of 15 can read compared to only 39% of the population in Mozambique.

Income and material style of life at the household level are often used to estimate household assets. Material style of life creates a scale of assets by looking at the presence or absence of household structure, such as whether mud or concrete was used to construct the house, and the types of amenities and appliances in the house, including TV, radio, and toilets. The types of infrastructure development in the community, such as schools, roads, medical facilities, and others can be used to create a community-scale index of relative well-being. These three measures are presented for our case study of fishing communities in the WIO (Fig. 6.6). Assets at both the household and community levels in Madagascar are the lowest in the region when compared to communities in Kenya, Tanzania, Mauritius, and the Seychelles (McClanahan et al. 2008b). Addressing basic needs, such as food and shelter security in Madagascar, may be a crucial first step before communities can begin to meaningfully engage in adaptive management and conservation (Marcus 2001, Cinner and Pollnac 2004).

Conclusions

Examining key components of social vulnerability to changes in the flow of goods and services revealed that there is considerable spatial variability in how WIO societies will be affected by, cope, or take advantage of opportunities provided by climate change. Importantly, national-scale measures of vulnerability may not necessarily reflect the local-scale situations in coastal societies. Also, there is considerable intra-country variation in key components of vulnerability. Strategies to reduce vulnerability will need to recognize those components of adaptive capacity that can be increased and aspects of sensitivity that can be reduced based on existing strengths and weaknesses in these communities. The following chapters continue exploring the social dimension of vulnerability, specifically the ways that societies respond to changes, developing indicators of adaptive capacity, and key strategies for reducing vulnerability in coastal societies.

7

Coastal Communities' Responses to Disturbance

Human history is filled with environmental change and subsequent societal changes that were a mix of successes and failures. Often environmental change spurs innovations, which can help societies to overcome challenges, as was described for the development of irrigation and hierarchical societies in Chapter 3. Alternatively, entire societies can collapse because they make poor decisions or delay actions and fail to adapt to environmental change (Ehrlich and Ehrlich 2004, Diamond 2005, Lambin 2007). The ability of societies to adapt to environment changes has social, economic, cultural, political, and technological roots.

In an environmental change context, adaptation is making adjustments that reduce vulnerability to climatic change (Smit et al. 2001). These adjustments can be behavioral, technological, economic, legal, or institutional. Adaptations can be reactive to past or current changes or in anticipation of future change (Smit et al. 2001). For example, early warning systems for excessive rainfall in semi-arid systems can provide 2 to 6 months of lead time before the onset of malaria outbreaks. This lead time can be used to develop interventions to reduce mortality (Boko et al. 2007). Adaptation can also occur on a range of levels—from individual behavior, to economic sectors, to the collective action of entire societies at national and international levels.

As we saw in Chapters 3–5, western Indian Ocean coastal ecosystems are often exposed to extreme or highly variable conditions. In coastal communities, these conditions can include factors such as seasonal trade winds, coastal erosion, drought, intense storms, or even occasional cyclones. If WIO communities have continually been exposed to environmental change and have managed to adapt thus far, why should we be concerned about whether they can adapt to future changes?

Experience with previous environmental change and disturbance is likely to help some communities, but three factors make projected rapid climate change considerably different from regular climatic variability. First, climate change is expected to alter the frequency of high-intensity events on an unprecedented scale (Chapter 3). Rare events such as coral bleaching and mortality are projected to become more frequent, and cyclones are expected to take new courses and affect previously unexposed areas. Second, rapid climate warming is expected to have profound effects on food security

(Funk and Brown 2009). For example, crop yield could diminish 50% by 2050 and as much as 90% by 2100 in some parts of Africa (Boko et al. 2007). A rainfall compilation by Funk and colleagues (2008) indicates these trends are now in effect for East Africa with dire consequences for people and wildlife (Western et al. 2009). This is likely to place increasing pressure on marine resources, creating demand that is disproportional to the population growth. Third, contemporary pressures such as overfishing and land-based pollution can interact with climate change, causing marine ecosystems such as coral reefs to undergo sudden and unexpected shifts to algal dominated systems, with consequence for fisheries production (Jackson 2001, Jackson et al. 2001, Pratchett et al. 2008).

The environmental change we are likely to see in marine ecosystems will be different in scale and scope from most of the previous changes in living memory and the experience of most coastal societies. This will present novel situations, and previous adaptations are unlikely to be sufficient to cope with the expected future climatic changes (Boko et al. 2007). Additionally, we should be concerned because not all historical adaptations to environmental change were necessarily successful. There is a great deal of archaeological evidence that coastal civilizations have collapsed in the past few millennia due to mix of drought and warfare (Horton and Midddleton 2000). More than ever, coastal communities will need to anticipate and prepare for climate change if they are to prosper. Our aim is to understand what makes societies good at adaptation and whether there are basic principles that will help people deal with change and reduce the chances of catastrophe or collapse.

Climate Change in Coastal Systems

Climate change will affect various aspects of coastal ecology and have social and economic consequences for coastal communities. The social and economic impacts of climate change to marine systems can be placed into three broad categories: (1) direct impacts from sea-level rise, cyclones, and coastal erosion; (2) indirect impacts to economic sectors such as marine-oriented tourism; and (3) fisheries. Climate change will also impact agriculture and water availability, but these topics are covered in a range of other publications and are not the focus of this book (Funk et al. 2008).

SEA-LEVEL RISE, CYCLONES, AND COASTAL EROSION

Vulnerability to sea-level rise, coastal erosion, and increased cyclonic activity will obviously be highest where populations and critical infrastructure are in low-lying coastal areas. In the western Indian Ocean, there are roughly 11,000 kilometers of coastal zone and over 60 million people who live near and are dependent on coastal resources (Table 6.3). Here, coastal populations represent between 8% and 100% of the country's populations. On the national scale, countries such as Kenya and Madagascar are considered less vulnerable because the majority of the population lives in the highlands; however, they do have sizable coastal populations (Coughanowr et al. 1995). The direct impacts of these threats to coastal communities will mainly be in the forms of (1) damage to coastal protection and other infrastructures, including ports, wharfs, and road networks; (2) inundation and storm-surge-related flooding of low-lying agricultural and settlement land; (3) increased sea-water intrusion into fresh groundwater; (4) encroachment of tidal waters into estuaries and river systems; (5) accelerated erosion of coastal areas, including key tourism development areas; and (6) losses to tourism, recreation, transportation, and non-monetary cultural resources and values (IPCC 2001, Nicholls and Lowe 2004, Elasha et al. 2006).

About a third of Africa's coastal infrastructure is in danger of inundation and coastal erosion due to increased sea levels of 15 to 95 cm by 2100 (IPCC 2001). For example, in Tanzania, a sea-level rise of 50 cm would inundate more than 2000 km^2 of land and create about $51 million in damage (UNEP 2006). In Kenya, a 1 m sea-level rise could cause $500 million of losses to key agricultural outputs, such as mango, cashew, nuts, and coconuts (IPCC 2001). The costs of adapting to sea-level rise in coastal countries could be as high as 5% to 10% of GDP, although losses could be as high as 14% of GDP if adaptive measures are not taken (Boko et al. 2007). Three major responses to sea-level rise have been presented by the IPCC: retreat, adapt, or defend (Boko et al. 2007). Adaptation options include (1) building sea walls, pillar housing, and raised foundations; (2) digging new wells and boreholes in response to saltwater intrusion; (3) delineating flood and erosion hazard areas; (4) eliminating the mining of sand and gravel for construction material; (5) reforesting sand dunes; and (6) controlling socioeconomic factors, such as overexploitation of coastal resources, population growth, and pollution (Elasha et al. 2006).

Indian Ocean island states could be particularly vulnerable to changes in the location, frequency, and intensity of cyclones (Boko et al. 2007). Places where there are dams or that have experienced destruction of natural barriers, such as reefs, mangroves, and dunes, will also be particularly vulnerable to coastal erosion. These include deltas, barrier islands, atolls, and other low-lying islands (Coughanowr et al. 1995). Beaches and dunes in Mauritius, Mozambique, and Tanzania are mined for minerals and construction resources (100,000 tons/year = 200 hectares of lagoon area; Fagoonee and Daby 1995). Degradation of coral reefs has caused significant coastal erosion in Kenya and Tanzania (Magadza 2000). Likewise, a dam along the Zambezi River (Zambezi Delta region) has caused coastal erosion, saltwater intrusion, and a loss of fisheries (Coughanowr et al. 1995).

MARINE-ORIENTED TOURISM

Marine-oriented tourism in the WIO is at particular risk from climate change. Besides threats to tourism-related infrastructure, climate-related events including coral bleaching may detrimentally affect dive tourism markets. The overall economic damages from the 1998 coral bleaching event in the Indian Ocean, for example, was an estimated loss of U.S.$3.5 billion in tourism revenue over a 20-year time frame (Westmacott et al. 2000b, Wilkinson 2000). This happens through two main mechanisms.

First, coral bleaching can create a loss of reef-related tourism revenue by reducing the attractiveness of a particular destination or even eliminating diving or snorkeling altogether (Westmacott et al. 2000c, Uyarra et al. 2005). For example, a bleaching event would influence dive site selection for more than 75% of divers surveyed in Kenya and Tanzania (Westmacott et al. 2000b). On a larger scale, severe bleaching may influence where people take holidays, potentially eroding national and regional tourism economies. A survey revealed that 19% of tourists in Zanzibar, Tanzania, and 30% of tourists in Mombasa, Kenya, would change their holiday destination if they were aware of severe coral bleaching (Westmacott et al. 2000b). Uyarra and colleagues (2005) found that 80% of tourists visiting Bonaire would be unwilling to return for the same price in the event of coral bleaching. Furthermore, Graham and colleagues (2000) concluded that the number of tourists visiting Palau in 1999 was 5% to 10% lower because of the previous year's coral bleaching event.

Second, coral bleaching can also influence tourist satisfaction with diving, snorkeling, and other reef-related activities, thereby reducing the value of a vacation destination and the willingness of tourists to return (Graham et al. 2000, Uyarra et al. 2005). Several studies have examined how coral bleaching can indirectly affect the economic value of tourism through a diminished tourism experience. In a survey conducted after a coral bleaching event, 47% of tourists surveyed in the Maldives considered the dead corals the most disappointing experience during their holiday (Westmacott et al. 2000b). A comparison of diving income in Kenya and Tanzania before and after the bleaching found that the recreational value of diving did not change (Westmacott et al. 2000b). The post-bleaching divers were, however, less experienced than the pre-bleaching divers, suggesting that more experienced divers may have chosen to dive in alternate locations where bleaching was less severe.

Countries most vulnerable to climate change effects on marine-related tourism include those with specialty dive markets. Key adaptive responses to the marine-oriented tourism market will include the following:

1. Strong management and marketing of perceived risk. A key determinant of tourist arrivals is the tourists' perception of the experience. Evaluations suggest that this may be less related to the

state of the ecosystem than to the perceptions of risk. For example, the number of newspaper reports on problems at a destination may play a larger role than the actual level of risks. A strong tourism body that can mediate and respond to negative media reports can manage these perceptions.

2. Diversification of products, market sources, and segments. Different market sources, usually countries of origin, and segments, such as backpacker, family travel, and executive, respond differently to products, prices, and risk perception. Diversifying the market sources and segments that a company or destination depends on can increase resilience to disturbances.
3. Adaptive pricing policies and regulatory structures, such as taxes or subsidies to dive or snorkel on reefs, may provide flexibility to rapidly adapt to disturbances and target different tourist markets.

FISHERIES

Climate change is likely to have six main effects on fisheries and fishing communities (Allison et al. 2009, Daw et al. 2009). These include (1) changes to fish yields, (2) changes to fish distribution, (3) damage to infrastructure, (4) effects on human health and safety, (5) changes to fish markets, and (6) changes to climate-driven policies that will regulate the fishing industry more tightly. A particular suite of potential adaptations may help fishers deal with each type of impact. Both the impacts and the potential adaptations have a high degree of inter-relatedness, and the same climate change mechanism may have multiple effects. For example, changes in migration patterns of fish may affect how many fish are caught and the safety of fishers traveling to catch fish. It is important to note that these effects are likely to be deeply intertwined with other non-climate-related factors, such as market conditions and the availability of fish on a wider scale (Glantz 1992).

Different adaptations will rely heavily on the specific aspects of adaptive capacity discussed in the previous chapter. For example, adaptations such as increasing effort or fishing power may rely primarily on assets to invest in new gear or vessels, while reducing effort may rely primarily on the flexibility to turn to other livelihood options. Of course, these aspects of adaptive capacity are not exclusive, and elements of flexibility are required to change fishing effort. Likewise, some assets are required to switch occupations. Governments and donors that want to successfully facilitate specific adaptations need to understand, identify, and build the primary aspects of adaptive capacity required for specific effects (Chapter 9).

Importantly, adaptive capacity is nested, occurring at multiple scales. Individuals, households, and communities may be adaptable, but if all are embedded in a state with disastrous policies and a constraining rather than an enabling hand, people may be unable to prosper under change. Our examples below and largely throughout the book focus primarily on local (i.e., individual, household, and community) scale adaptation and adaptive capacity. Consequently, a key assumption is that larger scale aspects of adaptive capacity are not constraining, although we certainly acknowledge that this is not always the case.

Some adaptations will be anticipatory and need to be initiated before expected impacts manifest, while others will be reactive to the change (Smit et al. 2001, Daw et al. 2009). Anticipatory adaptations include investments in disaster preparedness, including storm warning systems, and coastal modifications, such as sea walls to protect against storm surges. Likewise, some adaptations, such as diversifying livelihoods, will be the primary responsibility of private individuals and enterprises while other adaptations, such as providing public infrastructure and communicating storm and weather warnings to fishers, will be the responsibility of governments and donors (Smit et al. 2001, Tompkins and Adger 2005, Daw et al. 2009). Key climate change impacts to fisheries, likely adaptations, primary types of adaptive capacity required, whether the adaptation is reactive or anticipatory, and whether the primary responsibility is public or private are summarized in Table 7.1.

TABLE 7.1

Adaptations of specific fisheries to the impacts of climate change

Impact on fisheries	Potential adaptation measures	Main types of adaptive capacity required	Responsibility	Reactive/ Anticipatory
Changes to yield	Access higher value markets with same product (i.e., improve processing)	Assets, social organization	Public/ private	Either
	Increase effort or fishing power	Assets	Private	Either
	Reduce costs to increase efficiency	Assets, flexibility	Private	Either
	Change gear (to target different fish or habitat)	Assets, flexibility	Private	
	Reduce fishing effort (i.e., diversify livelihoods or rely on other existing income sources)	Flexibility	Private	Either
	Exit the fishery	Flexibility	Private	Either
Increased variability of yield	Diversify livelihood portfolio	Flexibility	Private	Either
	Insurance schemes, rotating credit	Social organization	Public/ private	Anticipatory
	Improved information and communication technologies (to locate fish and maximize market returns)	Assets, learning	Private/ public	Either

(continued)

TABLE 7.1
(Continued)

Impact on fisheries	Potential adaptation measures	Main types of adaptive capacity required	Responsibility	Reactive/ Anticipatory
Change in distribution of fisheries	Change location of fishing activities (i.e., migration, increased spatial mobility)	Flexibility, assets	Private/ public	Either
Damage to coastal infrastructure and coastal communities from flooding, cyclones, sea-level rise, and storm surges	Hard defenses	Assets	Public	Anticipatory
	Managed retreat/ accommodation	Assets, flexibility	Public	Anticipatory
	Rehabilitation and disaster response	Assets	Public	Reactive
	Integrated coastal management	Assets, social organization	Public	Anticipatory
	Early warning systems and education	Assets, social organization	Public	Anticipatory
Health and safety	Weather warning system	Assets, social organization	Public	Anticipatory
	Investment in improved vessel stability/safety	Assets, social organization	Private	Anticipatory
	Early warning systems for awareness of disease outbreaks (malaria, shellfish diseases)	Assets, social organization	Public	Anticipatory
Changes to profitability and shocks to markets	Diversification of markets and products	Flexibility	Private/ public	Either
	Information services for anticipation of price and market shocks	Capacity to learn, assets	Public	Anticipatory
	Alter fishing effort	Flexibility	Private	Either
Influx of new fishers to certain areas	Property rights	Social organization, capacity to learn	Public	Either

Each impact has a suite of potential adaptations that assist fishers to adapt or cope. Each adaptation relies on different elements of adaptive capacity, responsibility as either private individuals/enterprises or the public sector, and may be reactive or anticipatory.

Source: Adapted from Daw and colleagues (2009).

CHANGES TO THE YIELD FROM CAPTURE FISHERIES

Changes include overall yields and increased variability in catches. These may result from (1) declining abundance of target species, such as a decline in coral reef fish after a severe bleaching event, changes to reproductive patterns as a result of changing temperatures, or declines in pelagic species as a result of changes in oceanographic processes; (2) increased incidents of disease in high-value invertebrates such as lobsters (Glenn and Pugh 2006); (3) changed distribution of fish, as a result of changes to fish migration routes; and (4) an increase in the number of potential non-fishing days as a result of predicted increases in wind speed and frequency of storms (Allison et al. 2005). These changes will not always be negative and in some instances, local yields may increase. For example, Allison and colleagues (2005) project that flooding will increase fisheries landings in the flood plains of Asia.

At the individual scale, fishers can pursue several options when their catch declines. These include continuing fishing, stopping fishing either permanently or temporarily, changing where they fish, changing the gear they use, increasing effort to make up for lost catch, reducing costs to increase the efficiency of fishing operations, and accessing higher value markets. An example of the last includes targeting higher value markets, such as the live food fish or species that are sold in the aquarium trade. These products can fetch several times the price of fresh, dried, or frozen fish. A number of examples exist where fishers use these types of responses to deal with declining stocks. For example, in Lake Tanganyika, declines in yield and conflict have driven many fishers to utilize other lakes, including Lake Victoria (Allison et al. 2005).

Responses to higher variability in catch and markets can include technological improvements to provide fishers with information about fish locations, markets, prices, availability, product options, and navigation (FAO 2007). Improved communications will also be critical in reducing risks associated with increased weather variability and storms by providing early warning systems and timely weather forecasts (Allison et al. 2007). Responses to increased variability may also include diversifying livelihood portfolios to include non-fishing supplementary income and the development of insurance schemes or rotational credit arrangements to help attenuate variability. Adaptations that allow fishers to diversify their target species or their markets, including live food and aquarium fish trade, export fillets, local fresh consumption, and dried fish, may also help them to navigate the volatility of price fluctuations.

CHANGING DISTRIBUTION OF FISH STOCKS

Climate change will result in changes to the distribution of some fish stocks (Glantz 1992, Stenseth et al. 2002, Cheung et al. 2010). For example, tuna stocks in the western Pacific are likely to shift east in response to projected shifts in ocean temperatures, having profound effects on household and national-level revenues (Aaheim and Sygna 2000). Obviously, changing distributions will have variable effects. For example, Glantz (1992) noted that in the United States following an El Niño southern oscillation (ENSO) event, catches of salmon declined dramatically in California, Oregon, and Washington, but increased in Alaska.

Adaptations to deal with changing distribution of fish will primarily include relocating fishing activities, and this involves either migration or increasing the range of fishers. Geographers tend to categorize the reasons for migration into two broad types: "push factors" and "pull factors" (Adger 2000, Cinner 2009). Push factors can include deleterious conditions in the migrant's former home, such as population pressures, poverty, and resource scarcity. Pull factors relate to the attractiveness of a new location, including economic opportunity and marriage. Negative effects of climate change to some areas, particularly those that are already marginal, are likely to be a push factor that drives people to new areas. This is likely to result in an influx of new fishers to certain areas.

Changes in or disruptions to other economic sectors, such as agriculture and tourism, may also promote migration into fisheries. Push factors such as resource scarcity have been cited as the cause of "roving-bandit" migrants who decimate resources and then serially move on to new locations (Kramer et al. 2002, Berkes et al. 2006). As migrants shift to new areas or new fisheries, their desire to use near-shore fishing grounds is likely to become highly contested. These mobile roving-bandit fishers are already considered a globally significant problem for fisheries management (Berkes et al. 2006). Thus, adaptations to deal with climate change are likely to create secondary effects, such as increased pressures on resources in some locations.

Climate change adaptation measures may run contrary to some fisheries management goals that aim to localize fisheries through promoting property rights. In Kenya and Chile, recent legislation provides a form of property rights to coastal fishers essentially restricting their ability to fish in distant fishing grounds and simultaneously providing incentives for stewardship of local resources (Gelcich et al. 2008a, Signa and Tuda 2008). In Madagascar, only fishers in villages adjacent to marine park boundaries are allowed to fish in the park's multiple-use zones (Cinner et al. 2009b). But these fisheries management measures that seek to localize fisheries may undermine people's ability to respond adaptively to climate change by migrating. Allison and colleagues (2007) note that fishers in Lake Chilwa, Malawi, coped with continually fluctuating lake levels by adopting diverse livelihood strategies that incorporated migration. Yet policies designed to improve their welfare were actually eroding their adaptive capacity by requiring fishers to be full-time residents in order to access the fishery, thereby eliminating the adaptation created by mobility.

INFRASTRUCTURE

Rapid climate change and associated sea-level rise, cyclonic activity, and storm surge will increase the damage to critical fisheries infrastructure, including wharfs, shipyards, roads, and transportation networks. The increased frequency of extreme events may also result in increased loss of or damage to gear, such as nets, traps, and long lines. For example, hurricanes in the Caribbean have destroyed or damaged up to 34% of the fishing fleet in Antigua and Barbuda, cost $1.2 million in lost gear in Belize, and caused Jamaican fishermen to lose 90% of their fishing traps (Allison et al. 2005).

Adaptations such as coastal protection, modification, and relocation from high- to low-risk areas and coastal planning that avoids construction in high-risk areas will avoid damages and disasters. Solutions that aim to minimize impacts on the agricultural sector by flood control may be detrimental to fisheries (Allison et al. 2005). Adopting integrated coastal zone management that maintains or restores wetlands, vegetation in catchment areas, and mangroves can also address flooding and coastal erosion issues without adverse effects on fisheries. In some cases, these integrated coastal management adaptations improve livelihoods by increasing harvests of invertebrates, such as crabs and mollusks (Allison et al. 2007).

HUMAN HEALTH AND SAFETY

More variable weather patterns and the need for some fishers to travel farther to follow changing fish distributions will influence the health and safety of fishers and coastal communities. Disease outbreaks such as malaria and cholera and increased incidents of shellfish poisoning resulting from temperature-dependent phytoplankton blooms are also expected (Allison et al. 2005, Baschieri and Kovats 2010). Adaptations to deal with increased risks to human safety and health will include investments in health care; disease prevention, such as mosquito net distribution; safer boats; and technology. Fisher safety can be improved when information and communication technologies, such as location devices, navigation equipment, and weather warning systems, are fitted to existing networks, including mobile phones. Remote sensing data and early warning systems on temperature and rainfall conditions can be combined with communication networks to identify and verify conditions that may foster disease outbreaks, such as malaria, and identify phytoplankton blooms that can cause shellfish poisoning.

MARKETS

Climate change is likely to change fisheries markets profoundly (Macfayden and Allison 2009). In addition to changes in fisheries distribution, abundance, and composition, high fuel prices as a result of regulating carbon emissions, diminishing fossil fuel supplies, and a reduction of fuel subsidies will change the profitability of offshore fishing. Increased travel and movement of processing infrastructure needed to follow changes in fish distributions and production are expected to increase the costs and amount of fuel and ice consumed (Allison et al. 2005). Increased climate disturbances, such as cyclones, will more frequently disrupt fish market supplies and logistics. These factors combined with a higher demand for healthy protein and both population growth and affluence will increase demand and costs considerably beyond the natural productivity of the marine ecosystems.

Changes to fisheries markets will be complex and unpredictable. Rising fuel costs will potentially lower the profitability of many fishing operations. In addition, expected increases in demand for fish and reduced supplies will likely result in higher prices. Globalized markets will mean that reduced supplies of one species in one location will increase prices for similar species in other locations (Daw et al. 2009). Adaptations to deal with changing markets will include improved information technologies that will provide fishers with real-time information about prices, availability, the distribution and abundance of fish, and other aspects of global markets. Fishers may also respond to changes or shocks in the market by diversifying their livelihoods into non-fisheries sectors.

CLIMATE CHANGE-DRIVEN POLICIES

Many of the biggest climate-related changes to fisheries will come from policies that seek to protect resources from climate change shocks. For example, in 2004, the Great Barrier Reef Marine Park authority expanded the proportion of no-take fisheries closures from roughly 5% to 33% of the barrier reef. Climate change concerns largely drove this policy because fishing is predicted to make these reefs less resilient to climate disturbances. Additionally, the models used by fisheries managers are likely to start incorporating climate change elements, such as risk and exposure (Grafton 2010). Consequently, fishers may find themselves increasingly dealing with policies that require them to adapt, including more no-take areas, fisheries buy-outs, and restrictions on gear targeting climate-sensitive species (Chapter 10).

Fisher Responses to Reduced Yield

Dealing with climate change will require new adaptations as well as an improved understanding of existing capacities, knowledge, and practices of adapting to change (Coulthard 2008). Understanding how fishers respond to disturbances is important for fisheries management and for insights into their potential response to changing catches associated with climate change. Fishers' responses to hypothetical change are one way to better understand their behavior and coping mechanisms, despite the limitations set by evaluating hypothetical scenarios. A study of coastal fishers from 29 different fishing communities in the WIO asked fishers how they would respond to four hypothetical scenarios of sustained declines of 10%, 20%, 30%, and 50% in their catch (adapted from McClanahan et al. 2008b, Cinner et al. 2009e). Fishers' decisions to continue, adapt, or stop fishing depended, in part, on the severity of the disturbance. Predictably, across countries, small disturbances elicited either no response or a small one, whereas fishers are much more likely to respond to larger disturbances (Fig. 7.1).

Interestingly, certain responses tend to dominate in specific countries. These different responses may reflect broad cultural preferences; economic conditions, such as access to technologies or alternative occupations; national-scale governance issues, such as whether state policies are enabling or constraining certain adaptation options; or geographic issues including the availability or extent of reef or key fishing areas. For example, the dominant response in the Seychelles was to move location. The Seychelles has a large

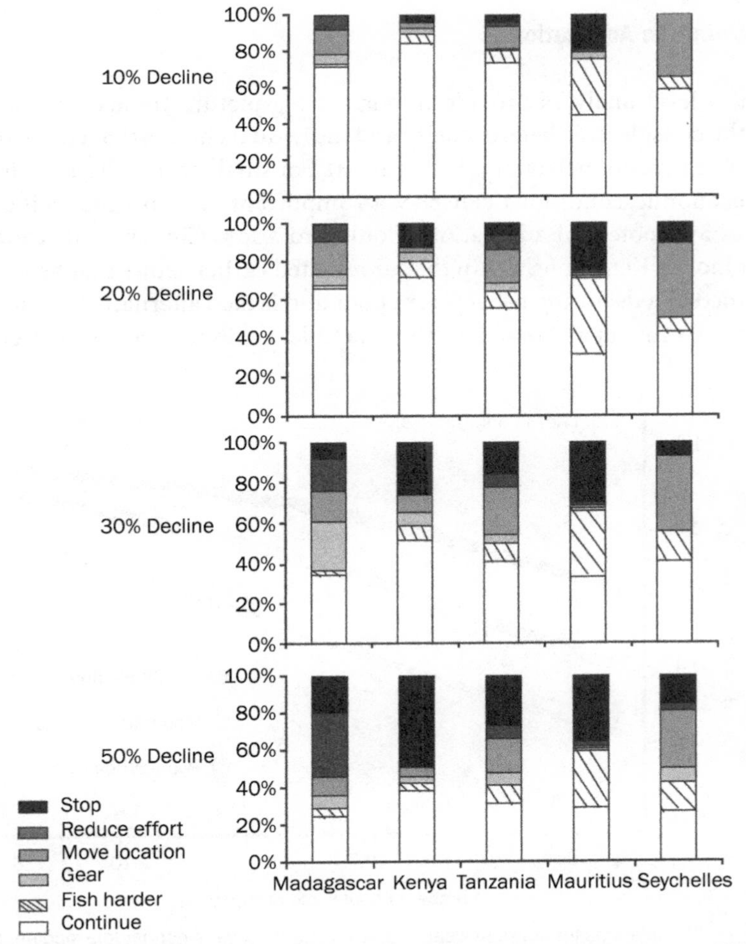

FIGURE 7.1 Fishers' responses to four hypothetical scenarios of declining catch rates (10%, 20%, 30%, and 50%). Results from Kenya, Tanzania, Madagascar, Seychelles, and Mauritius were examined separately.

Source: Adapted from Cinner and colleagues (2009a) and (2011).

plateau with abundant fishing grounds and Seychellois fishers tend to have boats with engines (Cinner et al. 2009c). The Seychellois response strategy reflects their high degree of spatial mobility. In Madagascar, the most frequent responses were to temporarily reduce effort. The relative abundance of land for agricultural activities and the high number of household jobs in this country could explain this choice (Cinner et al. 2009b). Interestingly, Mauritian fishers preferred the option of fishing harder, which may reflect their small reef area, low mobility, and fewer household jobs. Responses were clearly context specific and suggest that efforts to increase the adaptive capacity of fishers must recognize this context.

Constraints to Adaptation

Country-level analyses provide important big-picture trends but they can miss the considerable heterogeneity and individual choices that characterizes fishers' or investor behaviors (Wilen 2004). For small-scale fishers, household socioeconomic conditions can play an important role in influencing their choices and potential adaptation (Coulthard 2008, Cinner et al. 2009e). A closer look at Kenyan fishers in the survey cited earlier shows that fishers who remained in a declining fishery were poor and lacked alternative occupations (Fig. 7.2; Cinner et al. 2009e). In this example, flexibility and assets were two

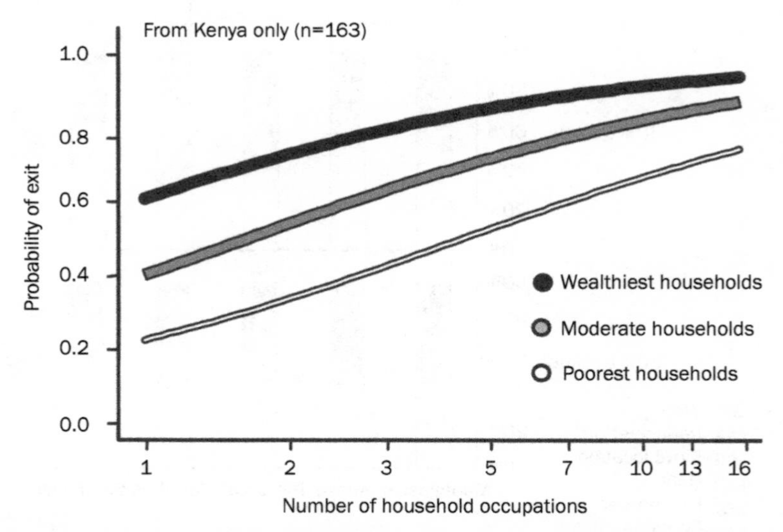

FIGURE 7.2 The relationship between wealth (divided into the wealthiest, middle income, and poorest), the number of household occupations, and probability that fishers would exit the fishery in response to a 50% catch decline. The lines show the relationships from a binomial logistic regression.

Source: Cinner and colleagues (2009a).

key aspects of adaptive capacity that influenced fishers' choices of how they thought they would respond to a declining fishery. A study of entry and exit from the Lake Victoria fishery also emphasized the role of flexibility, concluding that many continued fishing because of a lack of livelihood alternatives, or economic inflexibility (Ikiara and Odink 2000).

The role of both flexibility and assets in how people can cope with change is highlighted in a broad body of literature on "poverty traps," which are situations in which poor people are unable to amass sufficient resources to overcome shocks, such as natural disasters, or chronic low-income situations, such as a gradual decline in the fishery (Dasgupta 1997, Adato et al. 2006, Carter and Barrett 2006). Consequently, they are trapped in stable or increasing poverty because social exclusion and a lack of access to cash and credit prevents them from accessing higher income livelihoods (Dasgupta 1997, Adato et al. 2006, Carter and Barrett 2006).

In a poverty trap, the poor protect their few assets and are forced to choose livelihood strategies with low and short-term returns – even investments with declining returns (Dasgupta 1997, Barrett et al. 2006b). Poverty traps can also be reinforced because the poor are unable to take high-return risks that are more available to the wealthy (Barrett et al. 2006b). In a field study, Barrett and colleagues (2006b) found that the poor pursue lower risk livelihood strategies than wealthy households in both rural Kenya and Madagascar. The constrained flexibility of the poor makes them less able to adapt, but some of the mechanisms to escape this situation will be discussed in Chapter 9.

Maladaptations in Social-Ecological Systems

Certain adaptations are frequently viewed as positive because they help people to immediately cope with change, but they may have longer term ecological effects that erode the resilience of the wider social-ecological system. For example, the willingness and ability of fishers facing a decline in fisheries yields to choose a particular option can have direct effects on fish stocks. Adaptations to declining local stocks of marine resources, such as changing fishing grounds, using different gear, and fishing harder, have the potential to amplify the depletion of marine resources, particularly when many fishers respond the same way. Alternatively, reducing effort or stopping fishing reduces pressure on marine resources and allows fish stocks to recover (Cinner et al. 2011). Adaptations and adaptive capacity viewed from a long-term social-ecological system perspective may be different from adaptive capacity viewed from a purely individual or short-term social perspective.

An example of hypothetical responses to changes in the fishery, given by Tanzanian fishermen, illustrates how fishers' responses to declining yields have the potential to amplify a declining fishery (Cinner et al. 2011). As

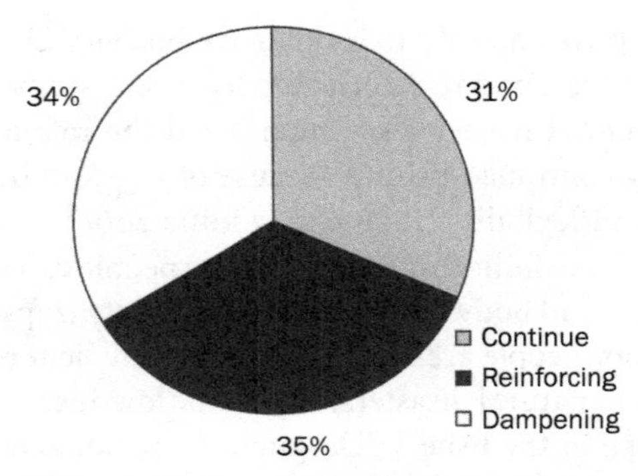

FIGURE 7.3 Tanzanian fishers' responses to a 50% decline in catch. Fishers' responses include these: continue fishing, reinforcing adaptation (such as changing fishing ground, switching gear, and fishing harder), and dampening adaptations (including reducing effort and stopping fishing).

Source: Adapted from Cinner et al. (in press).

described earlier, responses fall into three categories: dampening, amplifying, and continuing without change. Presented with a hypothetical 50% reduction in catch, slightly less than a third of Tanzanian fishers were likely continue without changes, a third would use dampening, and slightly more than a third would use amplifying responses (Fig. 7.3). Two factors would likely contribute to a general ratcheting-up effect on the fishery. First, about half of the fishers who planned to employ a dampening response with a 50% reduction in catch would first try an amplifying response with lower (10%, 20%, and 30%) catch reductions. Second, most fishers who would employ a dampening response would do so by taking a temporary break from fishing while engaging in another occupation but would start fishing again when stocks recovered. Many of the fishers who would use an amplifying response planned to add fishing capital, such as increased nets or traps during the period when catch was declining. This capital investment is likely to stay in the fishery. Thus, while responses that reduced effort were short term, responses that increased the intensity of the responses were longer term. This leads to difficulties in reducing effort and increasing stocks to a point where they can provide ecological resilience (Chapter 5).

Conclusions

Thus far, this book has laid out the key issues surrounding climate change in the western Indian Ocean area. It has explored the history of climate change, where it is likely to have the largest effects, how these effects are likely to influence both coastal ecosystems and societies, and the ways that people can adapt to these changes. This chapter highlighted the mechanisms whereby climate change will affect coastal societies and described the range of potential options fishers have for dealing with the various changes. Importantly, we saw how mobilizing these various options will depend on the adaptive capacity of individual fishers and the broader society. In cases such as the poverty trap in the Kenyan fishery illustration, fishers are unable to mobilize the necessary flexibility and assets to effectively adapt to change. In other cases, such as the Tanzanian example, these adaptations may create further stresses on the ecosystem that can erode its resilience. The following chapter develops the framework briefly introduced in Chapter 1 whereby key oceanographic, ecological, and social themes are integrated; the specific aspects of building adaptive capacity are discussed in Chapter 9.

8

Linking Social, Ecological, and Environmental Systems

The impacts of climate change in the western Indian Ocean (WIO) will be variable for both the marine ecosystems and the people that depend on them. Previous chapters described this variability and how it is influenced by the ocean environment that creates differing levels of exposure to heat stress and coral bleaching, the ecosystems, and the vulnerability of human communities. This variability, while complicating our understanding of its impacts, provides information that is critical for informing climate change policies and for developing appropriate human responses with the greatest chance of reducing ecological degradation and human suffering.

The heuristic framework we introduced in Chapter 1 to propose management actions based on specific combinations of social adaptive capacity, ecosystem status, and exposure to climate change is further developed here (Fig. 1.1). Quite simply, we consider each of these as a separate axis. When viewed simultaneously, the three axes form our framework, which creates a conceptual map or site niche. Based on a site's specific social, ecological, and environmental conditions, its location on this framework can be plotted. Specific combinations of social, ecological, and environmental conditions call for different strategies to confront the climate challenge. So knowing a site's "location" can provide some basic "directions" for how to get to or stay in more desirable place. We consider the most desirable location, or social-ecological state, to be one with low stressful environmental exposures, high social adaptive capacity, and an ecosystem that is able to provide important goods and services. As we shall see, the ways that societies can achieve this balance depends critically on the level of exposure to climate change disturbances.

Incorporating these social, environmental, and ecological dimensions into climate change planning can assist NGOs, management agencies, and donors to develop policy and management options that are more responsive to and reflective of local circumstances. We believe these more nuanced policies will have a better chance of being successfully implemented and adopted. Nevertheless, these three dimensions of environmental, ecological, and socioeconomic conditions are rarely considered together. For example, spatial differences in the status of ecosystems are frequently used to develop conservation and management priorities (Myers et al. 2000, Clark et al. 2001). However, this approach generally makes decisions based primarily on assumptions of environmental and ecological diversity and stability and either ignores or lightly weighs socioeconomic considerations.

We consider decisions based on these assumptions to be unrealistic, particularly given the risks imposed by climate change and the heterogeneous behavior of people. In some cases, two of these conditions have been simultaneously considered; for example, modeling payoffs for protecting high and low susceptibility sites based on their ecological conditions (Game et al. 2008b). This modeling approach advanced natural resource management theory by explicitly including climate risks, but it still did not evaluate the socioeconomic conditions that strongly influence the possibilities for management interventions. Social adaptive capacity is a critical consideration in the context of climate change adaptation and resource management because it is indicative of people's potential to act and adapt to change and take advantage of new opportunities.

In this chapter, we use a detailed regional case study of 29 coastal communities and associated reef systems across five WIO countries (Fig. 8.1) to

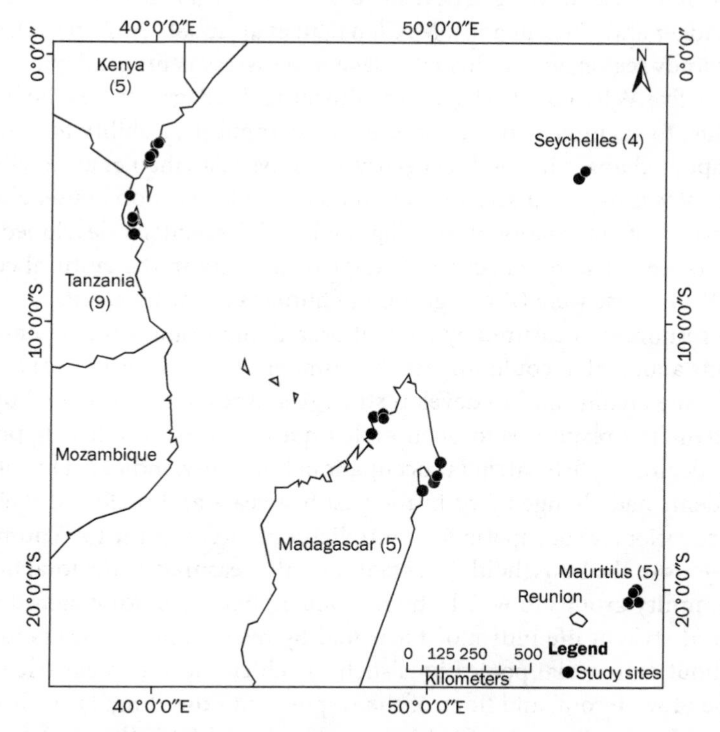

FIGURE 8.1 Map of study sites used in applying the framework across Kenya, Tanzania, Madagascar, Seychelles, and Mauritius.

illustrate how a measurable and normalized scale for each system (social, ecological, and environmental) can be developed and then plotted together to help inform adaptation and management strategies. Our case study and the indicators we use are specific to coral reef social-ecological systems and to the climate change threat of coral bleaching, but the general approach and framework are widely applicable to a broad range of systems and scales. The chapter has four parts: (1) development of a social adaptive capacity scale, (2) development of a scale of ecological conditions, (3) development of a scale of environmental exposure, and (4) operationalizing the framework with our regional case study. The following two chapters highlight specific strategies for moving systems along the social and ecological scales.

Social Adaptive Capacity Axis

As discussed in Chapter 6, we consider adaptive capacity a latent trait, which refers to the conditions that enable people to anticipate and respond to changes, minimize and recover from the consequences of change, and take advantage of new opportunities. Indices of social adaptive capacity to climate change impacts on fisheries have been developed on both the global and WIO regional scales (Adger and Vincent 2005, McClanahan et al. 2008b, Allison et al. 2009). For our study region, we conducted village-scale assessments of adaptive capacity across five WIO countries (Kenya, Tanzania, Madagascar, Seychelles, and Mauritius) to examine general patterns in a community's ability to anticipate and adapt to changes in coral reef ecosystems (McClanahan et al. 2008b). The assessment was based on a socioeconomic survey of over 1,500 households.

A panel of international and regional social scientists developed eight indicators we felt were reflective of adaptive capacity in the cultural context of the WIO. These were (1) recognition of humans as causal agents impacting marine resources, measured by content organizing responses to open-ended questions about what could impact the number of fish in the sea; (2) capacity to anticipate change and to develop strategies to respond, measured by content organizing responses to open-ended questions relating to a hypothetical 50% decline in fish catch; (3) occupational mobility, indicated by whether respondents had changed jobs in the past five years and preferred their current occupation; (4) occupational multiplicity, recorded as the total number of person-jobs in the household; (5) social capital, measured as the total number of community groups to which the respondent belonged; (6) material assets, a material style of life indicator measured by respondents' answers to questions about 15 material possessions such as vehicle, access to electricity, and the type of wall, roof, and floor of the respondent's dwelling; (7) technology, measured by the diversity of fishing gear used; and (8) infrastructure, measured by scaling 20 infrastructure items, such as hard-top roads, schools, and medical clinics (McClanahan et al. 2008b). Each of these eight indicators was normalized on a scale of 0–1.

The Analytic Hierarchy Process (AHP) methodology was used to aggregate the eight indicators into a scale of adaptive capacity. Ten researchers weighted each indicator based on its perceived importance to social adaptive capacity. An average of the researchers' weightings was used to calculate an adaptive capacity score for each community. The adaptive capacity for each site was calculated as follows:

$$\text{Adaptive capacity} = \text{Recognition of causality} \times 0.10 + \text{Change anticipation} \times 0.11 + \text{Occupational mobility} \times 0.11 + \text{Occupational multiplicity} \times 0.19 + \text{Social capital} \times 0.10 + \text{Material assets} \times 0.15 + \text{Technology} \times 0.13 + \text{Infrastructure} \times 0.12$$

Together, these indicators create an index that reflects people's ability to adapt to and take advantage of changes in coral reef resources. Importantly, these changes could be due to climate-related events, such as coral mortality, or changes to resource use resulting from policies, such as the closure of fisheries. A lower score on this index means people are more likely to be detrimentally affected by change.

Results of the adaptive capacity assessment indicate that communities in Kenya, Tanzania, and particularly Madagascar have the lowest aggregate scales of adaptive capacity and are likely to struggle with disruptions to the flow of ecosystem goods and services provided by coral reefs (Fig. 8.2). Despite broad national-level trends in adaptive capacity, there is a considerable spread within countries. For example, one site in Madagascar (Sahasoa) has higher adaptive capacity than many sites in Kenya, Tanzania, and even two Mauritian sites. Urbanized areas with higher levels of wealth and infrastructure also had higher levels of adaptive capacity. For example, despite lower national-level development rankings, such as the human development index (discussed in Chapter 6), the peri-urban sites in Kenya and Tanzania have similar adaptive capacity to some sites in the Seychelles and higher adaptive capacity than several Mauritian sites.

The intra-country heterogeneity suggests the need to consider adaptation policies within countries to account for differences in how specific communities can anticipate and cope with ecological changes. Importantly, there were a couple of sites (such as Sahasoa in Madagascar and Utange in Kenya) where there were data on adaptive capacity but no ecological data. Thus, we presented these to demonstrate the spread in adaptive capacity within countries, but a few sites will be absent in subsequent analyses, a situation that consequently tends to homogenize the within-country adaptive capacity.

Creating this additive and weighted measure of adaptive capacity is a first step in the development and testing of a local-scale measure of adaptive capacity. In the future it will be important to (1) test this measure against

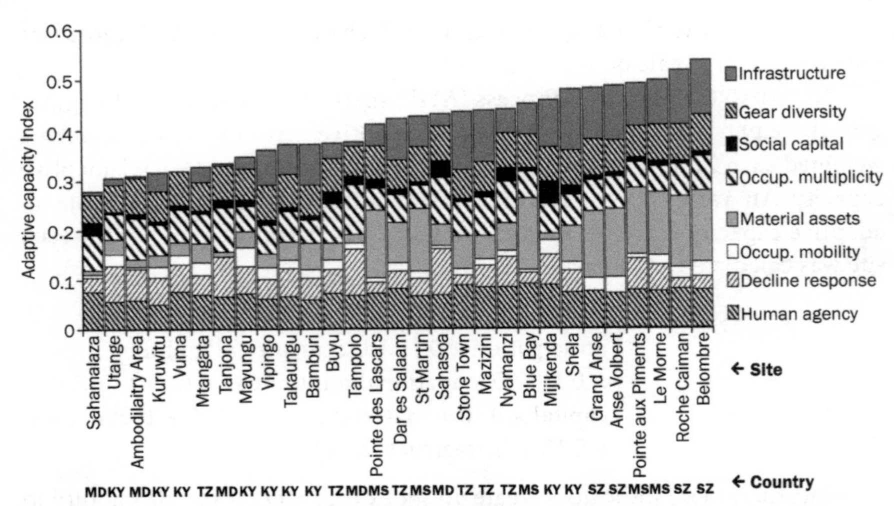

FIGURE 8.2 Index of the adaptive capacity of 29 communities measured as a compound of eight variables (right). MD - Madagascar, KY - Kenya, TZ - Tanzania, MS - Mauritius, SZ - Seychelles.
Source: McClanahan and colleagues (2008d).

actual responses to disturbance; (2) determine whether there are thresholds—for example, if one specific component of adaptive capacity is extremely low whether this will negate other aspects of adaptive capacity that are high; (3) test whether certain combinations of components are multiplicative (synergistic) or subtractive (antagonistic), rather than additive; and (4) better understand how national-scale issues, such as governance, influence this local-scale adaptive capacity.

Ecological Condition Axis

Like social systems, ecosystems are complex. Despite this complexity, indicators are frequently used to evaluate ecological conditions. For coral reefs, these are usually composed of key benthic cover elements, such as calcifying corals and algae, fish abundance, and the diversity of both groups. Many of these indicators are, of course, inter-related. The ecological component of our three-scale framework is composed of two fish and two coral indicators, which together provide critical information about the level of disturbance at a particular site.

Fish biomass and diversity are expected to be among the best indicators of fishing intensity or recovery after a reef is closed for fishing (McClanahan et al. 2009c). Fish biomass, which is the weight wet per unit area, has been shown to be a key variable for determining local status that is sensitive to management and human use (Cinner et al. 2009c). Fish biomass was calculated from standard field censuses in which the weights of all fish greater than 10 cm (fish large enough to be captured) were summed and put on a common area (i.e., hectare) basis. The number of fish species per 500 m^2 was used for four common and easily counted families: butterflyfish, parrotfish, surgeonfish, and wrasses. These families respond differently to fishing, habitat changes, and coral mortality; therefore, they represent both total diversity of the fish fauna and changes to disturbances (Wilson et al. 2006, Allen 2008).

The coral variables used to develop this axis were coral cover and the coral communities' susceptibility to bleaching. Coral cover is a frequently used indicator associated with local management (Selig and Bruno 2010) and larger scale disturbances, including temperature anomalies and coral bleaching (Graham et al. 2008b). The cover of live hard coral is expected to reflect mortality and recovery from coral bleaching, so sites with high coral cover are considered more pristine than those with low cover. The susceptibility measure is based on a large numbers of observations of bleaching among common corals during warm periods (McClanahan et al. 2007a). A site's bleaching susceptibility index is based on the relative density of each coral genus multiplied by the genus-specific response to bleaching and this taxa-specific response was summed across all taxa at the site to get a site metric (McClanahan et al. 2007a).

There are large differences in the response of coral taxa to warm water, and these differences are reflected in this susceptibility measure, showing that sites with low susceptibility are dominated by taxa tolerant of temperature anomalies, and vice versa. Coral reefs with high susceptibility indices have usually not experienced strong bleaching events, or they may have recovered quickly from these events. A more susceptible community has, in principle, been undisturbed by temperature anomalies and is composed of corals sensitive to climate and temperature disturbances, and therefore, is more pristine.

Similar to the social adaptive capacity as measure described, the two coral and two fish variables were normalized and then each variable was weighted based on expert knowledge of their sensitivity to or recovery from disturbance. The values were then combined into a common metric using the weighting technique explained earlier, which provides a 0–1 scale of ecological conditions for each site (Fig. 8.3). Different indices were explored and it was concluded that an index could be weighted toward corals, when primarily climate disturbances were being examined, and toward fish, when the focus was primarily local fisheries management disturbances (McClanahan et al. 2009d). The equation presented here is the one used for the coral weighted metric where

$$\text{Ecological conditions} = \text{coral bleaching susceptibility index} \times 0.35 + \text{coral cover} \times 0.35 + \text{fish species richness} \times 0.19 + \text{fish biomass} \times 0.10$$

Sites had considerable scatter along the ecological condition axis, but only nine of the 29 sites had values solidly above 0.5, most of which were fisheries

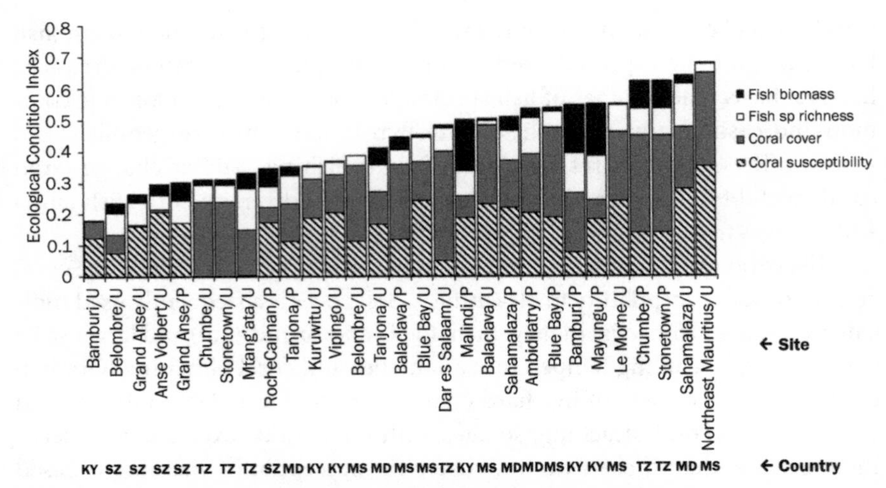

FIGURE 8.3 Index of ecological conditions for the studied coral reefs adjacent to the 29 communities where adaptive capacity was estimated.

Source: Adapted from McClanahan et al. (2009b).

closures with high coral cover, fish biomass, and numbers of species. The two sites with the highest values, however, were fished sites in Mauritius and Madagascar, which had coral communities with high coral cover and susceptibility. Sites with low values included fished sites in Kenya and Tanzania and all sites in the Seychelles, where reefs have been badly damaged by temperature anomalies.

Environmental Exposure

The risk of exposure to environmental disturbances provides critical information about the potential persistence of ecosystems under rapid climate change. Determining exactly when and where specific climate change impacts will occur is impossible, but there is information that can help to predict the probability and risk of a location's exposure to certain types of events. As described in Chapter 4, a number of oceanographic and environmental conditions can influence whether a site is likely to be bleached. These conditions, which are generally observable from satellite data, include sea-surface temperature (SST), maximum SST, the slope or rise of SST, ultraviolet radiation, wind speed, chlorophyll concentrations in seawater, photosynthetically active radiation, and zonal and meridional currents (Fig.3.5). Additionally, observations of coral bleaching have been consistently compiled throughout the regions (Fig. 4.7).

Combining this information allows us to map the exposure for corals and other organisms based on the known relationships between environmental variables and coral bleaching and the site-specific historical data on the oceanographic environment. The assumption of the model is that although the environment exhibits moderate inter-annual variability, there are predictable oceanographic patterns that influence rates of change and probability of extreme climate events (Webster et al. 2005, Baettig et al. 2007). If so, then stress models based on past histories can make useful predictions about future exposure and environmental risk. A model developed by Maina and colleagues (2008) weights 11 environmental variables described in Chapter 4 by their effect in increasing or reducing bleaching. It then combines all variables in the WIO to produce a map of exposure to bleaching (Fig. 8.4).

The map shows that the high exposure areas are in the north-central Indian Ocean, where most of the damaged and ecologically transformed reefs occurred (Ateweberhan and McClanahan 2010). From southern Kenya along the coastline there is an area of moderate exposure that still contains less disturbed or transformed reefs as well as the reefs from South Africa, eastern Madagascar, and Mauritius, which are predicted to have among the lowest exposures for the region. The Mozambique Channel and north is more variable ranging from moderate to high exposure.

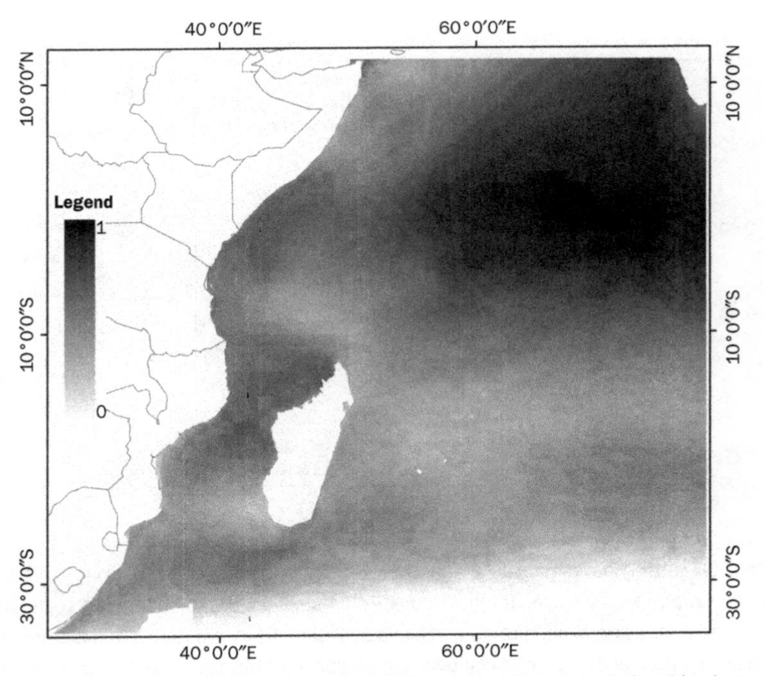

FIGURE 8.4 Map of the environmental stress model based on observed relationships between the environmental variables in Figure 4.3 and coral bleaching. Values are weighted based on measured effects. Darker areas are expected to have higher environmental stress.

Applying the framework

Of these three important considerations, only the social and ecological can be influenced by local actions. Ecological conditions can be degraded through abuses, including overfishing, or improved through management that reduces disturbances. Likewise, social adaptive capacity is something that can be built or eroded in a society. Interventions that can build adaptive capacity and ecosystem conditions are discussed in more detail in Chapters 9 and 10, respectively. Conversely, exposure to the effects of coral bleaching is based on factors such as oceanographic conditions, sea-surface temperature, and wind speed—variables that are not locally manageable. Of course, exposure can be influenced by global-scale actions such as reducing greenhouse gas emissions, which is the basis for the various global mitigation strategies, such as the Kyoto and Copenhagen agreements, but mitigation is a long-term program that goes beyond the scope of this book.

Here, we simplify the heuristic framework from Chapter 1 to provide a focus on the two components that can be locally influenced: adaptive capacity and ecosystem condition (Fig. 8.5). This creates four domains or quadrants of broad policy and actions. Plotting coastal communities in the WIO along

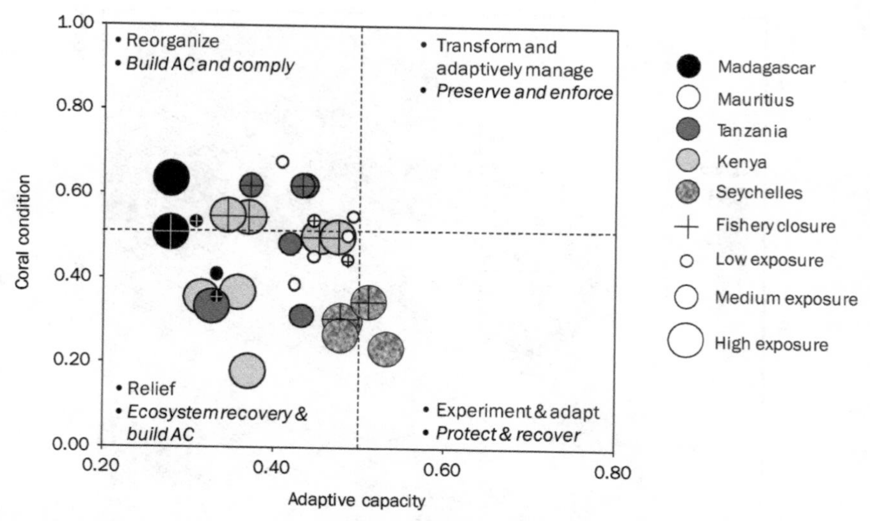

FIGURE 8.5 Scatter plots of the social adaptive capacity (AC) and ecological condition (weighted toward corals) scores for the 29 study communities. Broad adaptation goals will differ based on the intersection of AC and ecological conditions, creating four quadrants of influence. The level of exposure will influence more specific strategies. Specific adaptation strategies for low exposure sites are indicated in italic text and those for high exposure sites are indicated in regular text. The size of the bubbles indicates the site's exposure: small = low exposure and large = high exposure.

gradients of ecosystem condition and adaptive capacity indicates that these sites were scattered across three of the four quadrants.

Critically, more specific strategies also need to also be informed by the level of environmental exposure. In Figure 8.5, the level of exposure at a site is represented by the size of the bubble, corresponding to the level or exposure—larger means a higher level of exposure—and the specific strategies for high and low exposure sites are represented by the regular and italic bullet points, respectively. We separated study sites into the highest, intermediate, and lowest thirds of exposure values. Mauritius and eastern Madagascar had the lowest exposure, Tanzania had moderate exposure (except one site), and Kenya, the Seychelles, and western Madagascar had high exposure levels.

These differing levels of exposure and the ecological and adaptive capacity components have important implications for resource management. Even local-scale management can be informed by these exposure gradients. For example, some reefs in the high exposure areas will be damaged by climate regardless of successful management (Graham et al. 2008b). Sites with high exposure would, therefore, be poor targets for preservationist conservation activities that depend for their funding on consistent ecosystem quality to attract tourists. Reduced income from tourism after a bleaching event could potentially reduce the income needed for enforcement during the time that ecosystems are recovering and need the most protection. Supplemental funding would be needed to manage protected areas in these regions after bleaching mortality.

Protected area management that relies on user fees for management financing may need to target areas with low environmental exposure. Payoffs from conserving sites with high or low susceptibility depend on broader ecosystem conditions and recovery rates. In the western Indian Ocean, ecosystem conditions outside protected areas are generally degraded and often will not recover to undisturbed conditions. Therefore, an inference from previously discussed models is that payoffs for conservation will be greatest when protection is concentrated in low exposure areas and preferably among sites with high connectivity (Game et al. 2008b, Baskett et al. 2010).

None of the WIO study sites fall within the upper right quadrant with high adaptive capacity and good resource condition—the more desirable quadrant (Fig. 8.5). However, a number of sites border it. Importantly, sites are scaled along gradients of environmental, ecological, and social conditions. The quadrants in Figure 8.5 are guides to inform the types of policy actions that may best reflect local conditions. It is critical to note that the borders between the quadrants should not be viewed as distinct thresholds. Sites bordering a quadrant may require a combination of policy actions.

Most of the sites bordering this upper right quadrant are from Mauritius, but two fisheries closures from Kenya are also nearby. The Mauritius sites have low exposure, suggesting that these reefs are unlikely to be severely impacted by bleaching. Here, the resource should be more predictable and can be sustainably exploited with less risk of devastation from coral bleaching. However, this is also where the chances of maintaining an intact ecosystem and fishery are high and where we would propose strong restrictions on fishing effort and gear, and maintenance of high water-quality standards and large closures. Strong management is proposed because resource users are more likely to adapt and conservation will likely provide consistent benefits and revenue when the value of an intact ecosystem to tourism and national posterity is expected to be high.

The two high-exposure fisheries closures from Kenya bordering this top right quadrant will likely be degraded by climate change, as happened in the high-exposure sites in the Seychelles following the 1998 coral bleaching event. Our framework suggests that these areas will need to transform toward adaptive ecosystem management and lower societal dependence on these reefs. Supplemental funding mechanisms may be required to help manage these parks after bleaching events, which may decrease tourist satisfaction, numbers, and associated revenue. Likewise, the benefits local communities receive from the closure, such as increased income from fishing and tourism, will be expected to diminish over time in response to severe bleaching events (Westmacott et al. 2000a, Westmacott et al. 2000b, Graham et al. 2007). Consequently, the social-ecological system may need to transform away from dependence on these two fragile sectors. Adaptive management will be key to maximizing opportunities in both of these sectors (Chapter 9). Adaptive management will be critical in this sector to understand and respond to ecosystem change. Fishing methods outside parks should avoid fishing gear that targets species considered key to the recovery of coral reefs (Cinner et al. 2009a)—a strategy described in more detail in Chapter 10.

Sites in or bordering the low ecosystem condition but high-adaptive capacity state (bottom right quadrant in Fig. 8.5) have a mix of exposure values, suggesting that different strategies will be required to navigate the transition to improved ecosystem condition. Sites in the Seychelles are all in or nearby this bottom right quadrant. In these sites, improving ecosystem condition is a primary goal but must be done without undermining livelihoods or reducing adaptive capacity, which would be counter-productive. In the Seychelles, where exposure values are high, the road to recovery will be consistently set back by climatic events. Thus, integrating conservation with enterprises, such as eco-tourism, will be troubled by global competition with sites having lower exposure and better ecological conditions. High levels of exposure mean that improving ecosystem conditions may require experimentation and ecosystem engineering—activities that require high adaptive capacity but may ultimately inform and fuel innovations in other quadrants. Improving environmental conditions is expected to require harnessing fishers' capacity to adapt—for example, by sensibly shifting fishing capacity away from reefs to offshore pelagic resources. Additionally, here is where active ecosystem engineering may be experimentally used to facilitate ecosystem recovery.

Alternatively, Mauritian sites bordering this quadrant have low exposure and are most likely to be able to sustain reef quality in the long run if effective management policies are implemented. Here, fisheries closures or gear management strategies that integrate elements of tourism are most likely to work ecologically if compliance is good. Due to low exposure and high adaptive capacity values, Mauritius should be considered a high priority for conservation. Mauritius currently protects only 0.9% of its 870 km^2 of reefs from fishing; existing closures are small and lack effective enforcement, as evidenced by one closure site having lower resource conditions than control sites. To rectify this, Mauritians should seek to achieve greater ecosystem protection. As reefs with higher exposure become degraded, a highly effective system of fisheries closures in Mauritius could be a very significant tourist draw, providing considerable foreign exchange.

A number of sites in Kenya, Madagascar, and Tanzania fall solidly in the bottom left of our framework with both low adaptive capacity and poor ecosystem conditions. Here, the overall goals are to build adaptive capacity and resource conditions, but recommendations for doing so vary depending on levels of exposure. The policy recommendation is to build adaptive capacity and promote the recovery of resources where exposure is low. In this quadrant, only eastern Madagascar and Mauritius sites have low environmental exposure where extensive ecosystem recovery is most likely. In the Kenyan and Tanzanian sites, which have higher exposure, efforts to build adaptive capacity and ecosystem conditions will be challenged by climatic events. In these sites, relief efforts may be needed, particularly where changes to marine ecosystems coincide with changes in terrestrial ecosystems, such as decreased rainfall (Funk et al. 2008).

Early warning systems may be used to help prevent climatic events such as bleaching or drought from forcing people into poverty traps. Participation in the fishery will need to be balanced with other economic activities, and it is essential that adaptation and development strategies in these high exposure sites do not increase the dependence of local communities or industries on the reef-based resources that are at risk. Management measures such as creating protected areas, that can create profound changes to the flows of ecosystem goods and services, will be most difficult for resources users to accept if these users have low adaptive capacity or little ability to harness the benefits of a tourism-based economy. Here, management measures must focus on techniques with lower social displacement or those that help build resource users' social capacity to adapt to a different economy.

Sites in the top-left quadrant with high ecosystem conditions but low adaptive capacity are almost all protected areas. The sites in this quadrant displayed the range of exposure values. Critically, the goods and services that people obtain from these high-exposure protected areas are at risk from climate change.

Attempts to build adaptive capacity associated with flows of goods and services, such as through tourism or sustainable fisheries, are likely to suffer setbacks. Resource use and management strategies may need to be reorganized to account for these likely changes. Alternatively one Malagasy site and three Mauritian sites in this quadrant had low exposure and are likely to provide a more steady flow of ecosystem goods and services. Here, the focus is on maintaining compliance with protected area management while building adaptive capacity.

Other Applications

The framework previously described is an example of a general approach that can be applied to a wide range of social-ecological systems, scales, species, and stressors. For example, metrics of national-scale adaptive capacity to water stress, agricultural production, and models of projected exposure to drought could be used to examine national-scale food security in a region (Tompkins and Adger 2004, Funk et al. 2008). Alternative axes such as social sensitivity to climate change described in Chapter 6, the strength of governance systems, and the ability of ecosystems to provide important goods and services are other examples.

In many cases, investigators will be unable to collect the targeted and original data and will need to develop indices that are applicable based on the limits and scale of available data. A study by Allison and colleagues (2009) on the vulnerability of national fisheries to climate change is an example in which social and environmental aspects of climate vulnerability are examined based on widely available national-level data. The state of fisheries management at the national level is another example showing how the capacity of nations to manage their fisheries can be examined (Agnew et al. 2009, Alder et al. 2010). Many of these types of studies, however, lack the interdisciplinary context and often examine issues at the national scale.

In some cases, the weightings of specific variables may need to be adapted for particular scales of management. In the example used in this chapter, the ecosystem was weighted more heavily toward corals because the principal concern was the effects on them of climate change. Alternatively, if one is concerned with the state of the fisheries, then the fish variables could be given more weight (McClanahan et al. 2009d). Here, we illustrate this by developing a similar ecosystem condition equation, but with a higher weighting for fish biomass:

$$\text{Ecological conditions} = \text{coral bleaching susceptibility index} \times 0.20 + \text{coral cover} \times 0.12 + \text{fish species richness} \times 0.20 + \text{fish biomass} \times 0.49$$

Plotting coastal communities and reef conditions along this revised gradient of ecosystem condition and adaptive capacity produces some differences between the two calculations (Fig. 8.6). Most notably, the ecological

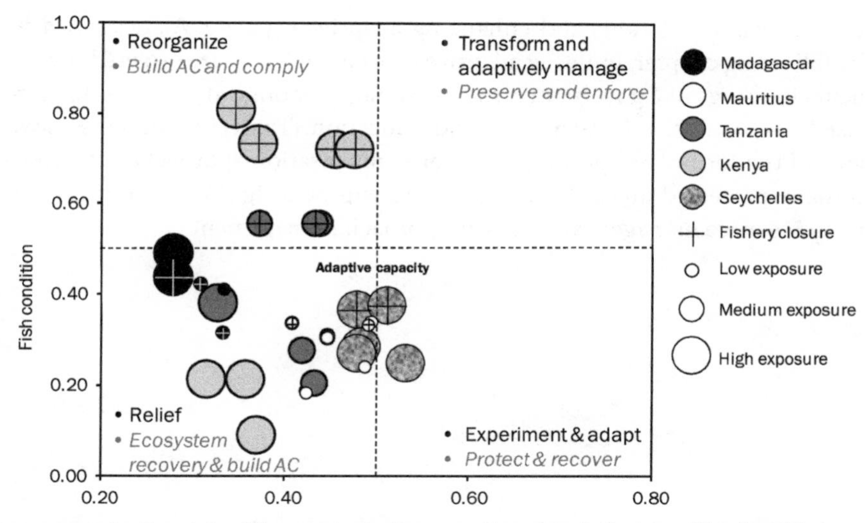

FIGURE 8.6 Scatter plots of the social adaptive capacity and ecological condition (weighted toward fish metrics) scores for the 29 study communities. The size of the bubbles indicates the site's exposure: small = low exposure and large = high exposure.

conditions of Kenyan fishery closures were raised considerably and sites in Kenyan fished reefs and other countries were often lowered by this metric. This revised metric, therefore, better reflects the status of fish and local-level management of the fisheries, compared to the coral-weighted metric. By the revised metric, a number of sites fall lower on the ecological condition axis and indicate a greater need to build or improve the site's fisheries and reef resources. The two methods combined produce insights into large-scale climate and smaller scale fisheries disturbances and associated insight useful for management at these two scales (McClanahan et al. 2009d).

Conclusions

The framework we developed here provides a basis for understanding some key local contextual conditions and then prioritizing pragmatic local-scale actions based on an integrated view of environmental exposure, ecosystem conditions, and human adaptive capacity. We believe that this integrated regional approach will be more likely to result in successful resource management in the context of climate change than trial and error, or policies and actions that place maximum value on either ecosystem protection or poverty alleviation. We also believe that the current emphasis on closing large fisheries to build ecological resilience and minimize the impacts of climate change is most likely to be successful in regions where high adaptive capacity and low exposure intersect. Other regions will require more focus on creating moderate resource-use restrictions and enhancing adaptive capacity. As described in the following chapter, building adaptive capacity will require targeted investments in things like governance, social capital, economic alternatives to reef-based livelihoods, infrastructure, and education. These investments move beyond the formulaic "participation" or compensation approaches common in many protected areas (Peters 1998) and will be a significant shift in how many resource managers and donors approach management.

9

Building Adaptive Capacity

Social adaptation occurs on various scales: some adaptations are undertaken by individuals or social groups, while others are undertaken by governments (Adger et al. 2005a). A list of recommendations for climate change adaptation on national and international scales, developed by the IPCC, includes these: (1) creating mitigation and adaptation policies, laws, and plans, including National Adaptation Plans of Action; (2) identifying and restructuring laws that hinder adaptation, including perverse subsidies; (3) coordinating action among organizations such as the Food and Agriculture Organization (FAO); (4) improving governance and transparency; and (5) developing mechanisms to handle trans-boundary adaptation issues, including regional warning systems, risk communication, emergency evacuation plans, and water resource management (Boko et al. 2007). The role of national- and international-level action is critical, as are efforts to integrate across social-ecological scales, but these topics are well covered in a number of publications (Adger et al. 2005b, Brooks et al. 2005, Boko et al. 2007). This chapter builds on these works by highlighting key crosscutting issues on how coastal societies can build adaptive capacity on more localized scales.

As discussed in Chapter 6, adaptive capacity refers to the conditions that enable people to (1) anticipate and respond to change, (2) minimize and recover from the consequences of change, and (3) take advantage of new opportunities arising from change (Adger and Vincent 2005, Grothmann and Patt 2005). This capacity is often considered a general trait, which means that someone with high levels of adaptive capacity is likely to be able to deal with a variety of shocks or changes. However, when the concept is used in practice, it is often done with a specific scale, shock, or change in mind. For example, in the previous chapter, we examined adaptive capacity at the household and community scales to changes in coral reef systems.

In this chapter, we discuss building general adaptive capacity, which we define as attempts to improve people's broad ability to cope with and adapt to change, and to take advantages of the opportunities provided by change. The following explores some key findings from the various academic disciplines that have long histories of application and research in building adaptive capacity, including human geography, development studies, agricultural economics, and understanding social-ecological systems.

We focus primarily on the individual, community, and local institution scales, and give examples related to natural resource use and management where possible.

Identifying Key Strengths and Weaknesses

On the local scale, practical attempts to enhance or build adaptive capacity often overlap with key components of sustainable development, such as promoting livelihood diversification, adding value to products, reducing poverty, improving literacy, and employing good governance (Smit et al. 2001). But not every community will need enhancement of the same aspects of adaptive capacity—some, for example, may already have diverse livelihood portfolios or effective governance, and attempts to further diversify or to strengthen governance may have low marginal returns, be futile, or in some cases, even undermine existing governance structures (Gelcich et al. 2006). Chapter 6 provided a simple conceptual framework for evaluating key components of adaptive capacity by breaking it down into four key dimensions: flexibility, assets, learning, and social organization. This categorization is helpful when examining potential strengths and weaknesses in the adaptive capacity of coastal communities.

For example, this framework was used to identify adaptive capacity in communities adjacent to Madagascar's marine protected areas (MPAs; Cinner et al. 2009b). Several aspects of local-level adaptive capacity appeared quite desirable in the Malagasy communities. In particular, flexibility in both livelihood strategies and the formal institutions governing marine resources provided some latent ability to adaptively use and manage marine resources. In general, these aspects of adaptive capacity needed the least attention. Conversely, several key dimensions of adaptive capacity were lacking, including household and community assets. Further, there was a generally poor understanding that humans were causal agents in marine ecosystems. People's inability to recognize that they influence marine resources coupled with their low levels of formal education potentially inhibited their ability to understand and incorporate scientific information generated from ecological monitoring programs in nearby marine parks.

In the Madagascar study, assets and learning were the aspects of adaptive capacity needing the most attention. Augmenting assets and helping people learn were expected to help them overcome short-term shocks created by changes in access to resources and to take advantage of the long-term opportunities resulting from the creation of the MPAs. Exceptions to these generalized strengths and weaknesses at the site level included less diverse livelihood portfolios near the Sahamalaza MPA and higher formal education at Nosy Antafana, which although low, was twice as high as in other, adjacent parks. These exceptions suggest that targeted strategies for building adaptive capacity are needed among communities. This example illustrates the need to identify strengths and weaknesses and work on the weaknesses with the highest potential returns.

Building the Four Components of Adaptive Capacity

BUILDING ASSETS

How poverty affects the ability of people to take advantage of opportunities—a critical aspect of adaptive capacity—is evaluated by a substantial body of literature. People can be trapped into behaviors that actually reinforce their own poverty because they are socially excluded, lack critical skills, and are unable to take the same risks as the wealthy. As we saw in Chapter 7, these situations are often referred to as poverty traps. Breaking poverty traps and the reinforcing behaviors will be a critical component of improving adaptive capacity in many parts of the WIO. Many poverty reduction projects have failed, however, and these efforts face many daunting challenges.

The pathways into and out of poverty are critical for understanding and building adaptive capacity, and a number of studies show the fluidity of this dynamic. Large-scale studies, such as an evaluation of post-apartheid KwaZulu-Natal, South Africa, found that 25% of the households surveyed fell below a defined poverty threshold, while only 10% moved out of poverty between 1993 and 1998 (Adato et al. 2006). Likewise, across 36 villages in Andhra Pradesh, India, 14% of households climbed above a poverty threshold over 25 years, but 12% of households fell into poverty (Krishna 2006). Similarly, a study in western Kenya, found that 19% of the households surveyed moved above a poverty threshold, but the same percentage fell into poverty over a 25-year period (Krishna et al. 2004). Importantly, poverty is often viewed as a multidimensional concept encompassing various aspects of well-being. Accordingly, these studies all used multiple poverty indicators, which included, for example, whether households had adequate food or clothing, provided education for their children, owned livestock, owned land, and so on. Both the indicators of poverty and the thresholds that determined whether someone was classified as poor were locally defined.

Households falling into poverty will be less effective at self-initiated adaptations (Adger et al. 2005a). In each of the evaluations, typically some combination of the following factors were responsible for people falling into poverty: (1) health issues, such as costly illnesses were the primary reason in Kenya; (2) sporadic high expenses, such as funerals and weddings; (3) large family size; (4) small landholdings; (5) personal debt caused by high-interest loans; (6) death of a key household member; and (7) environmental conditions, such as drought and fire.

In these studies, the accumulation of assets, diversification of livelihoods, and land improvements were factors most often related to people's ability to move out of poverty (Adato et al. 2006, Krishna 2006). In Kenya, diversification into the formal private sector accounted for 61% and government jobs for 13% of the upward mobility. Interestingly, in the Andhra Pradesh study, diversification into the formal private and government sectors accounted for very little upward mobility, just 9% and 14%, respectively. Diversification into informal entrepreneurial activities, such as small shops, or agricultural diversification accounted for more than 50% of their upward mobility. Irrigation and other government assistance projects were also associated with improved well-being, but these projects had mixed success, and the failure of some irrigation projects was cited as a key reason for the descent into poverty. Likewise, the efficacy of government assistance programs was patchy; some types of programs, such as loan assistance, did not work anywhere, while others, such as land improvements and new agricultural technologies, worked in only some locations.

In the South African study, those households that invested in and accumulated structural assets were more likely to remain out of poverty than those that did not, but the former were less than half of the households that emerged from poverty (Carter and May 2001). Consequently, more than half the households moved in and out of poverty by chance economic events, but those that invested in structural assets were less likely to return to poverty.

Building the asset component of adaptive capacity at the household scale will require two distinct types of policies and programs: those that prevent people from falling into poverty, and those that help people escape poverty (Barrett and Carter 2001, Krishna et al. 2004, Krishna 2006). The former involves investing in "social protection" programs that prevent a decline into poverty. These may include health care, family planning, and access to information and fair loans through social networks and mentoring. The latter will require policies that help the poor build assets beyond the poverty trap threshold such as the following:

1. Accessing new sources of productivity—for example, by focusing on activities with high returns or diversifying livelihoods. In the Kenyan artisanal coral reef fishery, programs might include gear-exchanges for those practicing illegal and destructive techniques, such as using beach seine nets.
2. Improving productivity of existing asset bases. This requires new technologies and improved efficiencies and, in the case of agriculture, improved water management by irrigation, storage, and efficient use, and soil conservation (Funk et al. 2008). Investments in basic infrastructure are needed to access markets and technologies, which can enable people to convert assets into income; but the wealth can

be short-lived when the productivity is not sustainable (Barrett et al. 2006a). In private-property agriculture, market access can provide money, diversification, and wealth, but the situation for fisheries is less straightforward because of the difficulties of managing common pool-resources and achieving fisheries sustainability. For example, a study in Nicaragua found that in the short term, the development of new roads led to the emergence of new products and markets and increased the export from fisheries (Schmitt and Kramer 2010). In Papua New Guinea and the Solomon Islands, market access has led to overexploitation of the fishery (Cinner and McClanahan 2006, Brewer et al. 2009). The biomass of fish can be removed and depleted quicker with improved access, creating temporary wealth, but this does not ensure sustainable use, which requires restrictions discussed in the next chapter.

3. Increasing opportunities to access capital, credit, and insurance. These are important in helping people protect investments in assets, but markets rarely deliver these services in a form that is affordable to the poor (Barrett and Carter 2001, Adato et al. 2006, Badjeck et al. 2010). Access to credit from financial institutions is difficult for many coastal people, particularly for fishers. Sometimes credit is provided by middlemen or people who purchase gear or fish but often at high interest rates and with incentives for heavy exploitation of marine resources (Fulanda et al. 2009, Crona et al. 2010). Thus, increasing microcredit opportunities and developing insurance systems for coastal communities should improve the maintenance of their assets.

On a larger scale, the international fisheries contracts discussed in Chapter 2 have the potential to help build assets at the national scale that can assist with fisheries regulation and monitoring. Currently, these are not implemented sufficiently to meet these goals but remain a potential mechanism for sustained financing.

BUILDING FLEXIBILITY

Livelihood diversity, one of the primary aspects of flexibility, was discussed in the preceding section because it is often closely linked with successful poverty reduction (Krishna 2006). For example, in North Sulawesi, Indonesia, the introduction of seaweed farming improved villagers' material style of life, a measure of assets (Sievanen et al. 2005). Diversification can also occur within fisheries if fishers are provided alternative gear and equipment. For example, rapid re-tooling of Peruvian fishing boats allowed fishers to increase their catch of shrimp during the 1997–98 El Niño (Badjeck et al. 2010). Importantly, diversification is not an option for all villagers because some are so deeply trapped in poverty that risking something new is unrealistic for them. Additional barriers to diversification include lack of skills, contacts, and access to capital—areas where the poor are often marginalized (Krishna et al. 2004, Boko et al. 2007).

The flexibility of organizations and institutions to adapt to change is another key dimension of adaptive capacity in the context of climate change. Adaptive management is considered an ongoing process by which organizational and institutional arrangements, rules for management, and ecological knowledge are continually monitored, tested, and revised in a process of learning-by-doing (Olsson et al. 2004). It is frequently assumed that institutions and organizations that adaptively manage resources will be more successful at confronting the challenges of climate change (Folke et al. 2005, Badjeck et al. 2010, McCook et al. 2010). We believe that institutions and organization must have some key components of adaptive capacity to adaptively manage resources. For example, adaptive management requires legislation that provides the flexibility necessary for these revisions to occur (Olsson et al. 2004, Folke et al. 2005).

The decentralization of marine resource management to local communities, as is happening in places including Kenya and Madagascar (Cinner et al. 2009d), has been critical in making institutions more flexible and adaptive. For example, following the development of the beach management unit (BMU) legislation in Kenya in 2006, there has been an interesting change in the governance of near-shore marine resources. In particular, a new system of small, community-managed closures, locally called *tengefu* (which means "set aside" in Swahili), has rapidly emerged along with interest in locally restricting the use of destructive fishing gear. Socioeconomic and ecological conditions in the BMUs are monitored through partnerships with international NGOs, and this information is expected to result in the refining or changing of rules over time.

By allowing resource users to develop and revise their own rules, the BMU legislation provides flexibility regarding the operational rules for managing fisheries. In some cases, this may allow for the reduction of effort that can help increase fish stocks. However, enhancing adaptive capacity at one scale may undermine aspects of adaptive capacity at another scale. Specifically, the BMU legislation may actually restrict other aspects of fishers' flexibility—particularly the ability of individual fishers to migrate in response to fluctuations in resources. The legislation allows BMUs to charge fishers from other areas for landing fish within the BMU, potentially creating barriers to mobility along the coast. In Chile, similar property rights legislation was developed that effectively restricted overexploitation caused by migrant fishers or roving bandits, but this also created a problem for those fishers who were not members of groups with property rights (Gelcich et al. 2006). Adaptation policies will need to consider and potentially resolve the inherent conflicts between property rights and human mobility. In some cases, such as when local and migrant fisher groups are targeting nonmigratory stocks, conflicts may be irreconcilable, but when targeting migratory or seasonal stocks there are better chances for resolution.

This concept of adaptive decentralized management is sometimes called "adaptive co-management" (Armitage et al. 2009). Decentralization as practiced throughout the WIO does not always provide the flexibility necessary for adaptive management. For example, some community-based marine resource management initiatives in Madagascar have had to operate outside the existing national governments' co-management framework because government rules were too rigid and did not allow for rapid change (Cinner et al. 2009d). A specific example is the Velondriake marine protected area where rotational fishery closures did not comply with the rigid geographic boundaries imposed by the government's co-management system. In Velondriake, reef areas are generally closed to fishing for 2–6 months to rebuild marine resources, particularly octopus. The system is truly adaptive in that the size, the length of closure period, and locations are influenced by users' perceptions regarding the improvements of harvests in previous closure attempts. This rotational closure system requires a more flexible framework that allows rules to be changed according to the perceived or measured results. Additionally, it is not just community or co-management institutions that must be adaptive; government agencies such as the Kenya Wildlife Service, which manages the largest and most successful marine reserves in the region, will need to become more adaptive as they learn about the exposure of their parks to climate change.

Building the flexibility components of adaptive capacity will require development of skills and access to capital to allow the most marginalized people to diversify their livelihood portfolios. Investments in technology may assist fishers in making rapid transitions during times of change, allowing them to take advantage of new opportunities (Badjeck et al. 2010). Importantly, programs promoting livelihood diversification or new technologies should not attempt to diversify into sectors that may be highly sensitive to climate impacts, particularly in locations with high levels of exposure to climate change.

Flexibility in institutions will be improved by supporting the emerging transition toward adaptive co-management, but these need investments and capacity building to ensure that management can respond to feedback from social and ecological systems. Also, specific adaptive management policies, such as BMUs, may limit other aspects of flexibility, such as the ability of fishers to migrate along the coast. It is important that property rights for marine resources that allow exclusion for fisheries management be developed to curb roving bandits but also allow for responsible migration and adaptation to environmental change (Berkes et al. 2006, Badjeck et al. 2010).

BUILDING THE TYPES OF SOCIAL ORGANIZATION THAT MAY FACILITATE ADAPTATION

National-scale changes to social organization can greatly influence how societies adapt to change. For example, as Vietnamese society transitioned from a centrally planned to a market economy, the capacity to develop centralized adaptation planning was limited; but the market economy provided the wealth that helped to reduce vulnerability for some, but not all, people (Adger 2000). The WIO region is generally characterized by weak national-level governance, which can profoundly influence local-level adaptive capacity. Addressing issues including corruption, transparency, and stability of national governments will be key to building the trust required for effective social organization and adaptive capacity at local community levels.

Building the capacity of people to network or to act collectively is another key part of fostering the type of social organization that helps people adapt to climate change (Adger 2003). For example, a global review found that in times of disaster, social cohesion helped reduce the loss and damage of assets, such as boats and coastal dwellings (Badjeck et al. 2010). Examples of efforts to build social capital include increased communication among communities and international NGOs as part of the establishment of MPA networks in Fiji (Sano 2008). International NGOs facilitated inter-community communication on resource management, which built social capital that reduced potential conflicts created by the establishment of MPAs. Another example from Trinidad and Tobago documented how a collaborative management process improved adaptive capacity by strengthening social networks among people responsible for disaster planning (Adger 2003). An example from western Kenya highlighted the importance of social networks and connections in allowing poor households to enter the urban economy (Krishna et al. 2004). Here, a social connection, such as a patron or family member, was critical in making the transition to the urban economy, whereas education, intelligence, and hard work alone were insufficient. There are, however, limits to what social capital can do, and the post-apartheid South African society is an example, where social capital helped poor households stabilize but not improve their well-being over time (Adato et al. 2006).

A key element of fostering cooperation between people is creating institutional incentives for people to work together. To help ensure cooperation, common property institutions are thought to benefit from specific institutional designs to help provide a credible commitment that resource users will follow the rules and effectively monitor these commitments (Ostrom 1990). Leading research on adaptive management, common property resources, and effective governance of complex social-ecological systems has found a number of common design characteristics shared by successful and long-enduring community-based management systems: (1) there must be clearly defined geographic boundaries and membership rights; (2) there must be congruence between rules and local conditions (i.e., scale and appropriateness); (3) resource users must have the right to make, enforce, and change the rules; (4) individuals affected by the rules must be allowed to participate in changing the rules; (5) the resources must be monitored; (6) there must be accountability mechanisms for those monitoring the rules; (7) sanctions must be enforced that increase with repeat offenses or severity of offenses; (8) conflict resolution mechanisms must be in place; and (9) institutions must be nested within other institutions operating on a larger scale (Ostrom 1990, Becker and Ostrom 1995, Ostrom et al. 1999, Ostrom 2000, Agrawal 2001). Box 9.1 examines whether these design principles are present in the co-management frameworks used in two WIO countries.

BOX 9.1

Design Principles in the Co-management Frameworks of Kenya and Madagascar

A study compared institutional frameworks for the co-management of marine resources in Kenya (called a beach management unit, or BMU) and Madagascar (Cinner et al. 2009d). Both co-management frameworks have many of the design principles that common property theorists have highlighted as important to the success of long-enduring co-management systems and were similar in the design principles that they lacked (see Exhibit 9.1). Here, we expanded Ostrom's (1990) principles to include monitoring of resources and monitoring of resource users. Monitoring is expected to provide information necessary for adaptive management.

Both systems have clearly defined membership rights, conflict resolution mechanisms, rights to organize, and congruence between rules and local conditions. In both cases, geographic boundaries are only partially clear because boundaries are often submerged landmarks, such as the edge of a reef. Likewise, collective choice arrangements are in place in both systems, but participation and input are limited to group members. For example, if a BMU decided to prohibit non–BMU members from access to its fishing grounds, these nonmembers would have no avenue for changing the rules. Neither country included graduated sanctions, such as warnings for first time offenders, fines for second offenders, and jail for repeat offenders, in its regulations.

Monitoring of the monitors was not conspicuously present in either system, but happened de facto in some cases. For example, the Kuruwitu community protected area in Kenya attempted to provide some transparency in the monitors by employing four monitors at any given time in the hope that peer pressure would decrease tendencies for corruption. The main components that were fully missing from these systems related to monitoring of resources and surveillance—aspects that are critical for adaptive management. In some instances, monitoring of the resources themselves were conducted by scientists and conservation groups, although key informants in both countries expressed concerns that the data and findings were not always made available to the concerned communities. In areas adjacent to marine national parks, park officials carried out enforcement, but in other areas communities were often left to enforce regulations themselves, without adequate training or equipment. This analysis of the design principles is a means to identify possible institutional gaps that may be the focal point for communities, NGOs, and other organizations.

EXHIBIT 9.1

Presence of design principles in co-management frameworks for Kenya and Madagascar

Design principle	Description	Kenya	Madagascar
Clearly defined membership rights	Clear delineation of membership rights to co-managed area	Yes	Yes
Congruence	Whether scale and scope of rules are appropriate for the local conditions	Yes	Yes
Rights to organize	Whether resource users have rights to make, enforce, and change the rules	Yes	Yes
Conflict resolution mechanisms	Rapid access to low-cost resolution forum	Yes	Yes
Nested enterprises	Nested within lead agencies or partner organizations at critical stages	(i.e. Partially)	(i.e. Partially)
Monitoring of monitors	Whether there are accountability mechanisms for those enforcing the rules	(i.e. Partially)	(i.e. Partially)
Clearly defined geographic boundaries	Clear delineation of co-management area	(i.e. Partially)	(i.e. Partially)
Collective choice arrangements	Whether individuals affected by the rules can participate in changing the rules	(i.e. Partially)	(i.e. Partially)
Graduated sanctions	Whether sanctions increase with numerous offenses or the severity of the offense	(i.e. Partially)	(i.e. Partially)
Monitoring of resources	Quantitative or qualitative monitoring of resource conditions	(i.e. Partially)	No
Monitoring of resource users (surveillance)	Whether appropriators' activities are monitored	No	No

Source: Cinner et al. (2009e).

Addressing design issues early on in the development of adaptive co-management institutions in these countries may enhance the chances of building robust institutions. Consequently, donors, NGOs, and governments should consider filling these potential gaps as a priority for institutional capacity building. Moreover, this should be done in a way that acknowledges the heterogeneity of institutions as critical and avoid adopting a one-size-fits-all policy for modeling institutions. Some of design principles may not be appropriate in a specific local context, and forcing them on local institutions, simply because they are recommended by theoreticians, can result in failure and alienation. Critics of the design principle approach have argued that a blueprint approach and lack of adequate consideration for important contextual factors critical for success can result from over-focusing on the internal characteristics of the institutions (Steins and Edwards 1999).

A strong institutional community structure is critical because weak decentralization can undermine resource management or adaptive planning if institutional capacity, coordination, and accountability are lacking and local communities act in the selfish interest of their leaders in isolation from larger community or national goals (Wickham et al. 2009). To address this, both political science and the social-ecological systems literature place considerable emphasis on those aspects of social organization that connect organizations at different scales—known as nested or polycentric institutions (Ostrom 1990, Armitage et al. 2009). The idea is that the capacity of societies to adapt is boosted by connections across scales because local organizations gain knowledge and strength by connecting with national or global organizations. Linkages with national or international organizations can help community-based management programs improve education, monitoring, and enforcement capacity. Alternatively, larger scale organizations gain local relevance and political support through their connections with local organizations. A specific type of organization, known as a bridging organization, can be used to connect organizations operating at different scales (Hahn et al. 2006).

How then can adaptive co-management organizations improve linkages across scales? One approach is to examine the hierarchy of organizations from global to local and their participation in the various stages of adaptive management cycle from rule initiation to modifying and changing rules (Cinner et al. 2009d; see Table 9.1). The following example describes participation in the emerging Kenyan BMUs where involvement is nested within a network of partners operating on a range of scales, from local to international. The Ministry of Fisheries Development, provincial administration, courts, and NGOs have clearly defined roles at different stages of the BMU framework, including initiating the development of rules, codification, enforcement, monitoring, and modifying, and making new rules. Improving linkages with agencies on different scales should be a priority for institutional strengthening in the BMU framework.

At the time of this evaluation, links at the international scale occurred only at select stages, possibly because of an antagonistic rather than a partnership relationship that developed in Kenya during an era when international donors were supporting NGOs in favor of the national government, which was perceived as corrupt. Competition for donor funds, resources, and control can weaken linkages but can be improved when these linkages become cooperative and participation occurs at more stages. We are not suggesting that each partner should be involved at every stage, but making sure there are consistent and supportive linkages across scales at all stages is expected to increase the adaptive capacity of these types of hierarchical organizations. Bridging organizations may be critical to helping BMUs better connect with national and international organizations. In the Kenyan BMU case, this is probably best played by the Fisheries Department, which already has representation at multiple scales of governance. The theory on the advantages of nested institutions assumes that the connections between scales of governance are low cost,

TABLE 9.1

Partners and organizational linkages at different scales for five different stages of BMU implementation and management in Kenya: (1) initiating the development of rules and regulations, (2) codification, (3) enforcement, (4) monitoring of resources, and (5) modifying or making new rules

Partner	Initiating rules	Codification	Enforcement	Monitoring	Modifying or making new rules
International					
International NGO/ donor				√	√
National					
Ministry of Fisheries Development	√	√	√	√	√
Courts of law			√		
Provincial					
Provincial administration	√	√	√		
District					
Ministry of Fisheries Development	√	√	√	√	√
NGO	√			√	√
Local community or landing site					
Management committee	√	√	√	√	√
Local users	√		√	√	√

Table is arranged with largest scale at the top and smallest at the bottom; processes in time are shown from left to right. BMU = beach management unit; NGO = nongovernmental organization.
Source: Cinner and colleagues (2009e).

transparent, and reinforcing (i.e., institutions are helping each other achieve common goals). However, when they are not, conflicts arise that can undermine successful coordination, create competition, and undermine productive social organization. Because the costs of interaction can be high, particularly where operational resources are scarce, and these conflicts frequently arise in the governance of common-pool resources, some observers of fisheries management have argued that single and accountable management organizations have the best records of success (Hilborn 2007).

Enhancing the social organization component of adaptive capacity in much of the WIO region will require the strengthening of regional coordination on cross-boundary issues, such as illegal fishing and migratory fish stocks, and making progress on national-scale issues of corruption, institutional harmonization, and effective governance. Local initiatives will also be critical to build social capital and support community-based institutions for adaptive co-management. Investments to build social networks and relations may increase cooperation and promote sharing of ideas, technologies, and adaptation strategies (Badjeck et al. 2010). Network and bridging organizations that coordinate and connect organizations at different scales will be expected to improve management effectiveness and enforcement capabilities.

The cost of building these organizational networks needs to carefully evaluate the economic basis and sustainability of the interactions, such that they persist after the termination of donor funding.

BUILDING THE CAPACITY TO LEARN

Fostering the ability of individuals and institutions to learn about climate change, to learn from a wide-range of experiences, and to learn to live with change and uncertainty is critical to building adaptive capacity (Adger et al. 2005a, Folke et al. 2005, Badjeck et al. 2010). Practical learning, traditional knowledge, formal education, and literacy are key components in people's ability to absorb and processes information on the causes and consequences of climate change.

Continuous access to information and opportunities needs to be part of ongoing learning and may be more important than formal education and literacy. For example, the previously discussed studies of poverty in Kenyan and Indian households found that formal education was not a significant factor in whether households moved above a poverty threshold over time (Krishna et al. 2004, Krishna 2006). Many educated villagers remained poor, in part because they lacked access to information and associated opportunities (Krishna et al. 2004). Organizations and institutional arrangements that provide timely and easily accessible information about resources and climate change will be critical to building adaptive capacity.

Information, such as seasonal forecasts, can help farmers choose crops with the best yields under new climatic conditions (Grothmann and Patt 2005). Early warning systems can help fishers assess potential risks, reduce lost or unproductive fishing days, and ultimately reduce deaths (Badjeck et al. 2010). In one of the very few long-term examples of building adaptive learning, a decade-long agricultural extension and conservation program in Zimbabwe developed tools to help farmers learn about bio-physical processes affecting crop productivity and promoted an experimental farming process that was highly successful in encouraging local adaptation (Hagmann and Chuma 2002). Programs that develop a set of practical tools for learning and experimentation and develop frequent feedback of information are, therefore, probably more important than formal education in building adaptive capacity.

Importantly, learning about climate change and adaptive management needs to be circular and iterative and not considered a one-way flow of information from professionals to societies. Combining different types of knowledge, including scientific and traditional ecological knowledge, can improve adaptive management (Berkes and Seixas 2006, Folke et al. 2005). For example, in the Roviana Lagoon in the Solomon Islands, local and scientific knowledge about the life history and preferred habitats of a key species, the bumphead parrotfish (*Bolbometopon muricatum*), were combined and used to protect the critical habitat (Aswani and Hamilton 2004). Likewise, creating and supporting networks that allow communities to share management experiences is one way to ensure a continuous information-building process. The Locally Managed Marine Areas network, which connects and shares experiences about fisheries closures among coastal communities across the Asian Pacific, is an example of this type of network (LMMA 2009). This type of support from a bridging organization can help share different resource and climate change knowledge.

Organizations that promote the capacity to learn will also be critical for developing adaptive capacity. These organizations should support education initiatives, including those focused on female education and reproductive health, helping fishers and farmers better understand climate change and adaptation options, combining local and scientific knowledge to inform adaptive management, and fostering experimentation with new resource capture techniques and other potential adaptations, such as seed crops. The learning component of adaptive capacity will also require that scientists and funding agencies develop communication strategies for dispersing comprehensible scientific information to communities in culturally appropriate means. Learning institutions will also need to look for ways to combine scientific and local knowledge systems (Aswani and Hamilton 2004, Berkes and Seixas 2006).

In much of the WIO, investments in formal and informal education for children and adults are necessary to increase the capacity to learn about climate change processes and potential disturbances to livelihoods. This will require investments in education infrastructure and capacity building for teachers and administrators (Peters 1998) but can be equally cost-effective by directing specific low-cost education at communities. Additionally, increasing female education is known to have strong links to key aspects of human development, including lower infant mortality and fertility rates (Browne and Barrett 1991).

Reducing Other Aspects of Vulnerability

This chapter highlighted key aspects and practical examples of how to build adaptive capacity, with a specific focus on the fishery sector. Of course, building adaptive capacity outside the fishery sector will be critical to reducing vulnerability to other shocks and adverse trends (Adger et al. 2005c). This may be achieved through adequate investments in health services and education as well as strengthening local governance and accountability mechanisms.

In addition, reducing other aspects of social vulnerability, including exposure and sensitivity, will help societies cope with climate change.

Lessening societal dependence and maintaining the productivity of reef fisheries will primarily reduce sensitivity to changes in coral reef ecosystems. For those that still fish, reducing sensitivity will be aided by changing to gear that does not select and unsustainably exploit fish most affected by climate change (Cinner et al. 2009a). Reducing sensitivity to the effects of climate change will, however, also require adaptations to agricultural systems, such as increasing reservoir storage capacity and planting crops that can withstand higher temperatures or more climate variability (Adger et al. 2005a, Funk et al. 2008).

Changing the exposure of specific reef systems to climate change disturbances may be impossible, but better information about exposure, such as when and where bleaching is occurring, can help managers implement adaptive management. Following bleaching, coral mortality typically occurs within the ensuing 20 to 70 days depending on coral species, and algae can quickly colonize the space previously occupied by corals (McClanahan et al. 2001). Managers and resource users may prevent algal overgrowth by encouraging grazing and reducing nutrients. For example, in the time leading up to and after a bleaching event, managers and resource users may temporarily ban fishing or the specific gear that targets herbivorous fishes most likely to graze the reef and prevent it from being overgrown by algae (Cinner et al. 2009a). These management options will be discussed in more detail in the following chapter.

Reducing societal exposure to climate change disturbances, such as high-intensity storms and sea-level rise, will require more traditional mitigation techniques, including the development of setbacks, vegetation barriers, seawalls, and other defenses. However, disaster preparedness and planning for change will also play a critical role. Specifically, controlling settlement in low-lying areas will reduce flooding disasters, and when dwellings are built there, the ground floor should be floodable (Adger et al. 2005a). Reducing exposure and sensitivity were only briefly touched on here because these concepts are covered in considerable detail in other publications (Adger et al. 2005a, Boko et al. 2007).

Interactions, Unintended Consequences, and Constraints

Adaptation measures can potentially create unintended and unforeseen consequences on other social and natural systems, creating uncertainty about the outcomes of adaptation activities (Adger et al. 2005a, Badjeck et al. 2010). For example, because there are few barriers to entering the fishery in Kenya, raising the wealth of fishers in isolation from other economic sectors may attract new fishers seeking this wealth, ultimately increasing harvesting pressure and leading to reduced catch per person and higher poverty (Allison and Ellis 2001, Cinner et al. 2009c). Additionally, interventions that aim to improve fishers' assets in isolation from other elements of adaptive capacity have the potential to create negative ecological feedbacks. For example, wealthier Tanzanian fishermen tended to respond to scenarios of declining resources by increasing effort and changing fishing gear, which are choices expected to amplify negative feedback in the ecosystem more than the choices made by poor fishers (Cinner et al. 2011). In this example, increased occupational flexibility was associated with choices more likely to dampen ecological feedbacks, such as reducing effort or exiting the fishery. Thus, attempts to build these fishers' adaptive capacity should consider interventions that foster ecological dampening responses, such as supplemental livelihoods, coupled with regulations that restrict choices that can potentially increase or amplify negative ecological impacts.

Another example of potential interactions between social and ecological dynamics comes from a large-scale study of social drivers of resource conditions across the WIO (Cinner et al. 2009c). Across 19 sites in five countries, the relationship between community-level socioeconomic development (one of the measures of adaptive capacity in Chapter 6 and 8) and fish biomass (one of the indicators in the ecological conditions axis in Chapter 8) was u-shaped (Box 9.1). Thus, the heaviest overfishing on coral reefs occurred in countries part way up the socioeconomic development ladder. Countries with either very low or very high levels of socioeconomic development tended to have reef fisheries in fair shape—containing about four times the reef fish biomass of the intermediate development sites. Critically, this study suggests that relationships between certain aspects of adaptive capacity and environmental conditions are not always positively associated and, at certain stages, building specific components of adaptive capacity may actually foster conditions that facilitate overexploitation. Strategies to build adaptive capacity in places with the lowest levels of socioeconomic development should be coupled with management policies that avoid the practices driving ecological conditions to the bottom of the u-shaped curve (Box 9.2).

Strategies that coordinate and plan for change across sectors, such as fisheries and agriculture, are more likely to identify and minimize unintended consequences of climate change adaptation (Badjeck et al. 2010). Additionally, planning for adaptation across scales will be critical (Adger et al. 2005a). It is not entirely clear, however, how adaptive capacity at one scale inhibits or enables adaptive capacity at other scales. As we have seen, the relationship can sometimes be inhibiting, such that policies or programs that build adaptive capacity at one scale actually undermine aspects of adaptive capacity at other scales.

There are important social and institutional constraints to reducing people's vulnerability that can include cultural and psychological factors that create limits to some adaptations. For example, farmers in Zimbabwe did not

BOX 9.2

Social Drivers of Ecosystem Conditions

The u-shaped relationship between economic affluence and local environmental conditions is referred to as the environmental Kuznets curve. The expectation behind this model is that at early stages of economic development, environmental conditions decline as societies become wealthier. As societies become sufficiently affluent and begin to demand environmental quality, increasing affluence eventually leads to improving environmental conditions (Arrow et al. 1995, York et al. 2003, Clausen and York 2008a,b, Lantz and Martinez-Espineira 2008). A combination of potential mechanisms could explain why the condition of reef fisheries in the western Indian Ocean has a u-shaped relationship with socioeconomic development. These include changes in the techniques people use to harvest fish (called the technique effect), the attributes or composition of economies (called the composition effect), and the scale at which people operate (called the scale effect). A parallel theory of ecological modernization suggests that it is not economic development per se that leads to improvements in environmental conditions, but rather the accompanying institutional changes, such as investments in scientific and management organizations (Clausen and York 2008a,b).

At the lowest levels of development, which were in Madagascar, traditional fishing methods do not allow high catches, and customary institutions, such as taboos, restrict fishing effort (Display 9.1, Cinner 2007, Cinner et al. 2009c). Where communities have slightly more economic development, fishing effort and fishing techniques are more effective and destructive, but fishers often lack the most expensive technology to fish off shore, which confines their efforts to local coral reefs. In these middle development sites, traditional socio-cultural institutions formerly used to govern resources can break down (Cinner and Aswani 2007, Cinner et al. 2007); enforcement is often lax when this happens because governments are weak and contemporary fisheries management measures ineffective. For example, in Kenya, which has some sites among the poorest fishery conditions, these customary institutions were once widespread but have largely broken down in recent years in favor of national fisheries development policies, laws, and programs that encourage open access (McClanahan et al. 1997). Unrestrained and destructive fishing techniques are practiced and the degraded conditions are reinforced by a poverty trap, whereby declining coral reef fisheries are exploited by the poorest fishers despite diminishing returns (Cinner et al. 2009e).

In sites with high levels of socioeconomic development, it is possible that low dependence on fisheries, effective central government policies, and technology may help to reduce pressure on coral reef fisheries. The banning and effective enforcement of destructive fishing gear are examples of policy responses associated with increased economic development. In the Seychelles, for example, policies promoting the sustainability of coral reef fisheries include the effective banning of such destructive devices as spear guns and beach seine nets. In Kenya, where these nets have been banned through co-management activities, there is evidence that catches have increased (McClanahan et al. 2008c). Likewise technological advances associated with increased development enhance the capacity of fisheries, including boats that allow fishers to move away from near-shore coral reef fisheries. These technique, scale, and composition effects probably interact with institutions in ways that produce the observed u-shaped relationship between socioeconomic development and reef fish biomass.

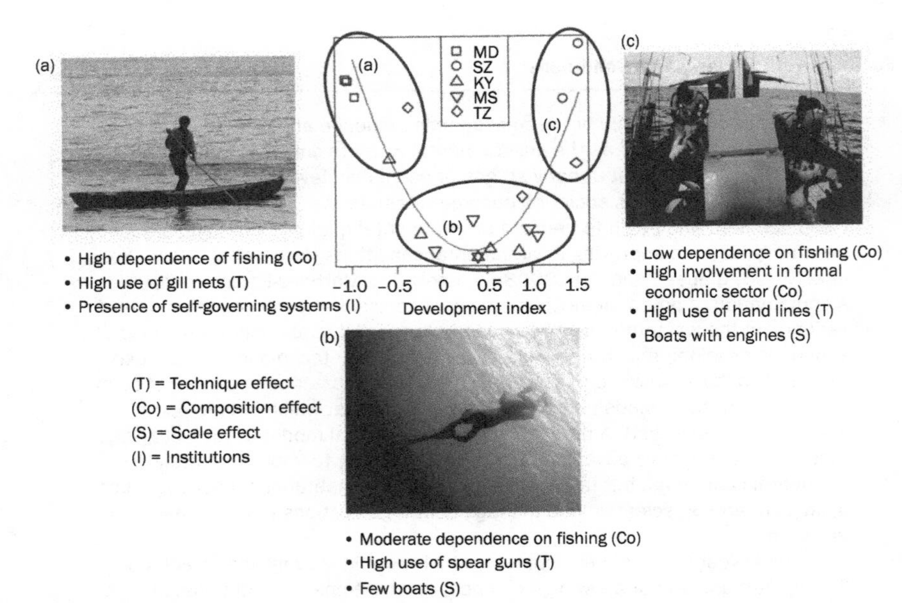

DISPLAY 9.1 The u-shaped relationship between socioeconomic development and reef fish biomass in 19 sites across 5 WIO countries. Cinner and colleagues (2009c) found a strong quadratic relationship between a multivariate index of community-level socioeconomic development and reef fish biomass ($R^2 = 0.77$, $p<0.001$). The socioeconomic development index was based on the presence or absence of 16 community-level infrastructure items such as schools, roads, and hospitals. The causal mechanisms behind this type of relationship can broadly be grouped into (1) the technique effect, (2) the composition effect, (3) the scale effect, and (4) institutions. Indicators of these causal mechanisms were significantly different among communities with (a) lower, (b) intermediate, and (c) higher socioeconomic development. MD = Madagascar, SZ = Seychelles, KY = Kenya, MS = Mauritius, TZ = Tanzania.
Source: Cinner et al. (2009c).

utilize localized climate forecasts to adapt their planting behavior because they perceived that all forecasts were equally likely and they did not believe their actions could avoid the possibility of failure (Grothmann and Patt 2005). Institutional constraints can include sociocultural institutions, such as taboos and constraining caste or ethnic systems, perverse incentives, unacceptably long gaps between implementation and realization of benefits, and time frames that do not correspond to political cycles, to name just a few.

Attempts to build adaptive capacity also need to consider issues of social equity in adaptation measures, such as who will win and lose, the social and legal legitimacy of adaptation processes, and the efficiency and costs of adaptation (Adger et al. 2005a, Thomas 2005, Paavola and Adger 2006). These are important considerations because certain adaptation measures, such as rebuilding infrastructure after a disaster, can exacerbate existing inequalities—making already vulnerable people even more vulnerable (Adger et al. 2005a, Thomas 2005). In general, building infrastructure to reduce sensitivity and exposure is expected to increase costs and unintended consequences more than building adaptive capacity. Understanding and overcoming constraints and inequities will be key to improving the efficacy of projects aimed to build adaptive capacity.

Conclusion

This chapter explored various strategies for building adaptive capacity, with a specific focus on the fisheries sector. Many of the policies and programs to build adaptive capacity will require external donor pressure and aid, as well as government support. Adaptation-specific funds such as the United Nations Framework Convention on Climate Change (UNFCCC) do exist to help governments implement their country's priority adaptations (Hope 2009). Additionally, international fisheries contracts discussed in Chapter 2 have the potential to support the building of adaptive capacity and sustainability around fisheries. International investments in low-income communities have dwindled over the last decade as donor development aid is increasingly being replaced by emergency relief and tied to macroeconomic policy conditions, such as economic liberalization, which frequently do not immediately favor the poor (Barrett and Carter 2001). International aid to fisheries is low, but it has increased in recent years with the recognition of the vast extent of this market and the increasing pressure to support sustainable fisheries. Donor and development agencies need to commit substantial aid to efforts that will build adaptive capacity throughout the hierarchy of resource use and management but particularly in the most vulnerable segments of society.

10

Managing Ecosystems for Change

Historically, fisheries management has focused on a limited number of species that were thought to operate in a stable or predictable environment. Management attempted to optimize maximum yields and profits through technocratic controls (Gordon 1954, May et al. 1979). Optimizing the harvest of target fisheries is a seductive goal that can occasionally work but is often challenged by (1) the complexity and multispecies nature of many fisheries; (2) erosion of critical ecosystem processes through overharvesting; (3) the continually changing nature of many marine ecosystems; and (4) the resource users themselves, who may have different incentives, needs, and aspirations from those of managers (Charles 1992). These realities create challenges that require appropriate management concepts and tools.

Contemporary approaches to fisheries management often attempt to incorporate change, complexity, and the human dimensions to maintain critical ecosystem functions and processes. As we saw in Chapter 5, ecosystems can have thresholds where changes occur that influence key processes. When pushed beyond thresholds, coral reef ecosystems can undergo what is known as an ecological phase shift and become dominated by algae, sea urchins, or other key species (Hughes 1994, Bellwood et al. 2004). The severe depletion or loss of species that play a key functional role in maintaining processes, for example, might make it impossible to return the ecosystem to some previous and desired state (Ives and Carpenter 2007). The concept of managing ecosystems to maintain the functions and processes that prevent them from undergoing a phase shift, is thought of as maintaining or building the resilience of the system—that is, its ability to absorb shocks, recover after disturbances, and resist phase shifts (Bellwood et al. 2004).

The broad purpose of this chapter is to highlight key strategies for managing fisheries in ways that build the resilience of coral reef social-ecological systems. Much of the science behind both historic and contemporary fisheries management has been generated in a wealthy country context. In many cases, the approaches developed are inappropriately applied to poor countries. In this chapter, we focus largely on the science and lessons learned from tropical fisheries management. Below, we identify critical thresholds in western Indian Ocean (WIO) coral reefs and discuss a range of management tools that can help prevent resources from falling below these thresholds. Although contemporary fisheries management tools can extend up the value chain to include market regulations, trade policies, and consumer choice, among others; for this book, we limit our discussion to managing the capture process.

Identifying Alternative States and Thresholds

Evaluations of coral reef ecosystems indicate that the state and ecology of coral reefs can fundamentally change along gradients of fishing (Dulvy et al. 2004, Cinner and McClanahan 2006, Mora 2008). Intensive fishing can promote the abundance of various algae, including erect brown algae, or sea urchins. This is because some fishing gear directly targets herbivores that feed on algae, and other gear targets the fish that feed on sea urchins (Cinner et al. 2009a). When these key functional groups are missing from reef systems, algae or sea urchins, respectively, are likely to dominate the system. The probability that these alternate ecological states will occur increases as the biomass of fish declines. There also appear to be thresholds of fishing intensity beyond which ecological change is significant and difficult to reverse. Identifying these threshold levels is key to developing guidelines for managing fishing impacts.

To identify the levels of fishing where thresholds occur, a novel statistical technique called switch-point analysis was applied to empirical data from about 330 WIO sites where similar ecological data were collected (McClanahan et al. in press). The switch-point analysis estimated the location of a change in the different ecosystem states where key ecosystem conditions and processes are believed to fundamentally change as fishing effort increases and the biomass of fish declines (Fig. 10.1). The analysis shows that the first switch point occurred at a fishable biomass of 1,130 kg/ha (hectare) where there was a rapid increase in the variability of erect macroalgae cover. This point was then followed by the ratio of macroalgae to living hard coral at 850 kg/ha and predation rates on tethered sea urchins at 640 kg/ha. The thresholds for six other metrics fell below 300kg/ha, which included fish species richness, sea urchin biomass, the proportion of herbivorous fish in the fishable biomass, calcifying benthic organisms, the ratio of sea urchins to herbivorous fish, and hard coral cover.

Consequently, various algae and sea urchins dominate moderately to heavily fished reefs. Fishing pressure and reduced fish biomass cause these declines in the numbers of fish species and the ecological process of calcification or reef building by influencing the abundance of key species, functional groups, and ecological processes. Coral cover is highly variable, likely due to the impact of the 1998 ENSO-driven coral mortality event in the region (Ateweberhan and McClanahan 2010) but coral cover also declined when fishable biomass reached the very low fish biomass level of 90 kg/ha. There is considerable scatter in the data and differences within countries, but these

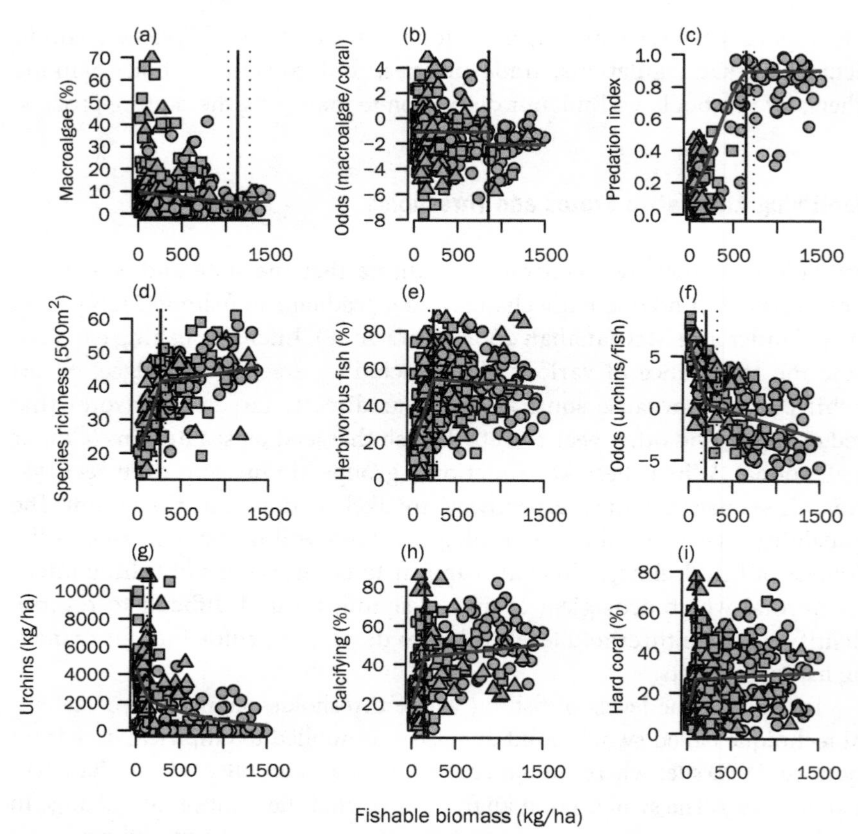

FIGURE 10.1 Ecological response to changing reef fishable biomass for nine ecological indicator metrics among reefs closed to fishing (circles), having open-access fishing (squares), and having gear-restricted fishing (triangles) in roughly 330 sites in the Indian Ocean. A. percentage cover of erect macroalgae; B. log-odds (1:1) macroalgae to coral; C. urchin predation index; D. numbers of fish species; E. proportion of herbivorous fish in total fishable biomass F. log-odds (1:10) of urchins to herbivorous fish; G. urchin biomass (kg/ha); H. percentage cover of calcifying substrates; I. percentage cover of hard coral.

Source: McClanahan and Colleagues (in press)

patterns are evident across this spectrum of fish biomass in many countries and can be used to develop management policies that avoid reducing fish biomass beneath key thresholds.

We have indicated where these thresholds lie within a generic conceptual model of the effects of fishing on marine ecosystems (Fig. 10.2, adapted from Worm and colleagues 2009). As the rate of exploitation increases along the *x*-axis, erect macroalgae and its ratio to coral abundance are the first metrics to change rapidly. As the rate of fisheries exploitation further increases, thresholds are reached for the predation rate on sea urchins (predation index), the number of fish species (species richness), ratio of sea urchins to herbivorous

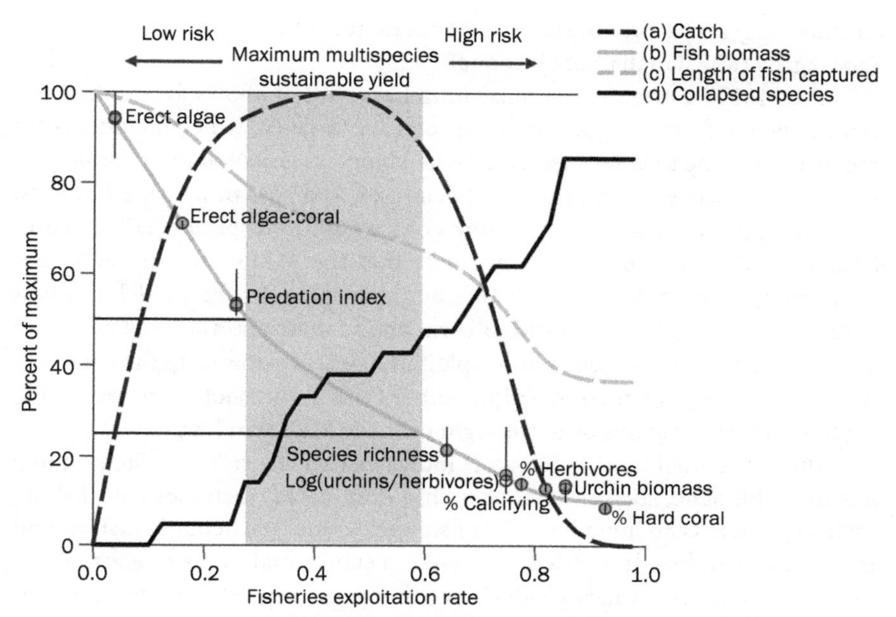

FIGURE 10.2 A model of changes to coral reef fisheries yields at different levels of fisheries exploitation based on a conceptual model from Worm and colleagues (2009) and threshold changes determined from data collected from coral reefs in the western Indian Ocean (McClanahan and colleagues in press). The model plots the relative measure of fisheries exploitation against (a) catch, (b) fish biomass, (c) length of fish captured, and (d) number of collapsed fisheries species. Percent maximum of a–d is on the *y*-axis. We plotted key ecological thresholds identified in our field studies along the relative fish biomass line, which ranges from pristine biomass (B_0) of 1200kg/ha (100% on *y* axis) to 0. We also indicate the window of MMSY (0.25–.5 B_0) and highlight with a gray box where this intersects with the relative fisheries exploitation rate on the *x*-axis.

fish, cover of calcifying organisms, percentage of fish biomass that are herbivores, sea urchin biomass, and hard coral cover, respectively. Consequently, changes in erect algae and predation on sea urchins are early warning signs for WIO reef ecosystems. Continued intensive fishing that further reduces fish biomass is likely to eventually result in a rapid loss of fish diversity, increases in sea urchins, and a decline in corals and other calcifying benthic organisms.

To help provide "goal posts" for resource users and managers, we bring together two fisheries concepts into this model; pristine biomass and multispecies maximum sustainable yield (MMSY). Pristine biomass is a metric for evaluating the condition of multispecies fisheries that is estimated from either historical catch data or from areas that have little to no fishing pressure (Hilborn and Stokes 2010). In the WIO, pristine biomass has been estimated by evaluating large, well-established fisheries closures (those greater than 5 km^2 and 15 years old) and very lightly fished reefs in the Maldives.

Results suggest that pristine biomass for coral reef fishes in the region is about 1200 kg/ha (McClanahan et al. 2009c).

The MMSY estimates the maximum harvest that allows for stocks to be replenished in fisheries where many species are targeted, which includes coral reef fisheries. The MMSY is expected when biomass is 40% of the pristine biomass, but estimates have ranged between 25% and 50% of pristine biomass, depending on the recruitment patterns of the captured species (Hilborn and Stokes 2010). Consequently, we estimate that the MMSY for the WIO lies between 25% and 50% of the pristine biomass value. In the model, the first three thresholds (algae, its ratio to coral abundance, and the predation rate on sea urchins) are reached at low exploitation levels below or to the left of the MMSY window (Fig. 10.2). All of the other critical thresholds are reached at exploitation levels above or to the right of the MMSY window.

The conceptual model (Fig. 10.2) indicates that the risk of switching to a less desirable state, such as algal or urchin dominated, increases with fishing intensity. There is an inherent risk in fishing near and particularly just beyond the MMSY window. In particular, there is a substantial range of effort at the top of the yield curve where yields do not change greatly, yet the risks of species collapse and ecosystem change increase considerably (Fig. 10.2). A lower risk strategy would be to fish at a harvesting level that maintains biomass within the window of MMSY, which is at exploitation levels below many of the key threshold points that are difficult to reverse. The lowest risk strategy would be to fish well below or to the left of this MMSY window. Next, we take a look at some of the types of management tools that may make this possible.

Fisheries Management Tools

The previous chapter discussed efforts to move the adaptive capacity axis, and here we do something similar for the ecosystem axis by discussing some of the key tools available to resource users and managers for improving the condition of reef fisheries. Management of the fisheries capture process can be categorized into six main restrictive actions. These include restrictions on the area where people can fish (such as closures), the times that people are allowed to fish (seasons), the size of targeted fish (minimum or maximum), the types of species targeted, the gear used, and the amount of fishing effort. These should all be seen as tools that can be applied in the appropriate social-ecological conditions.

RESTRICTIONS ON SPACE (CLOSURES)

Fisheries closures (often called marine parks, reserves, no-take, or marine protected areas) can provide a number of important ecological and social services, including the rebuilding of depleted species, providing sources of larvae, improving fisheries in surrounding areas, encouraging tourism, and reducing conflicts over resource allocation (McClanahan et al. 2006, McClanahan et al. 2007b, Lester et al. 2009, Pitcher and Lam 2010). Closures aim to protect the integrity of the wider ecosystem, as opposed to a single species. Consequently, this management tool is a precautionary approach that has received considerable attention from scientists, conservation groups, and managers.

Global reviews of published studies on the impacts of closures to fishes have found that closures can increase the abundance, size, and diversity of fishes (Lester et al. 2009), but it takes a number of years for the ecological benefits of fisheries closures to accrue. An evaluation of a number of the oldest and the best-studied closures in the world concluded that after fishing has ceased in an area, it takes about 5 years to detect changes in the biomass of target fish species and 13 years to detect effects on other taxa (Babcock et al. 2010). Studies of the high-compliance closures in Kenya indicate that full recovery of fish biomass stocks within the closure takes up to 22 years (McClanahan et al. 2007b). But closure size and the level of compliance also play a critical role in the benefits a protected area can provide. An empirical evaluation of 20 closures in six countries within the WIO found that the level of compliance, age of the closure, and its size were the key factors in determining the ecological benefits provided by the reserve (McClanahan et al. 2009c). However, the biomass of fish in closures increases greatly up to about 5 km^2, but above that size there was no additional increase in the biomass (McClanahan et al. 2009c).

Closures are thought to provide benefits to fisheries through a mechanism known as adult "spillover," whereby fishes from within the closure swim out into adjacent areas that are open to fishing. However, this will occur only when fishing effort is high and catches are below MMSY or on the left side of the fisheries yield hump in Figure 10.2. Although most studies find that closures increase the catch per unit effort in adjacent fisheries (Halpern et al. 2010), the response to total catch on a per area basis is more variable and is probably driven by aspects of closure design, such as closure size (Worm et al. 2006). Fisheries management in areas adjacent to closures also appears to influence the degree to which spillover happens. Gear and other restrictions adjacent to closures have been shown to result in increased fish biomass inside the closures and an extended distance of adult spillover from closure borders (McClanahan and Mangi 2000, Claudet et al. 2008). An illustrative example of this comes from the Mombasa Marine Park, where beach seine nets are used on the northern side and are effectively restricted on the southern side of the closure. Catches in the trap fishery were higher on the side where beach seine netting was excluded, but that this relationship was truncated on the side of the park where there was seine netting (McClanahan and

Mangi 2000). Spillover on the side that restricted seine net use was limited to the first few kilometers from the border, which is consistent with similar studies of spillover (Halpern et al. 2010).

Another aspect of spillover relates to the larvae from protected fishes or corals "seeding" outside areas. When closures are close enough to each other that a sufficient number of larvae can be exchanged, they are "connected." For fishes, this distance is thought to be on the order of tens of kilometers (Jones et al. 2009, Planes et al. 2009). Globally, there is considerable emphasis on making networks of closures connected through this type of larval spillover because such connectivity is thought to contribute to the resilience of protected populations, the overall network of closures, and the broader seascape (Jones et al. 2009). An evaluation of connectedness found that 34% and 49% of marine protected areas were connected by less than 20 km or 150 km, respectively (Wood et al. 2008).

There has been increasing interest among scientists and managers in examining whether unfished areas can buffer the impacts of climate change disturbances (Mumby et al. 2007, Knowlton and Jackson 2008). The rationale is that a high abundance of herbivorous fishes in unfished areas may also help prevent the buildup of erect algae on coral reefs following a bleaching event, in turn leading to improved coral recruitment, survival, and recovery rates (Mumby et al. 2007). Some studies indicate that remote, vast, and relatively pristine seascapes are resilient to a wide range of disturbances (Sandin et al. 2008, Sheppard et al. 2008). There is also some evidence that coral recovery is higher in older closures (Selig and Bruno 2010). However, there is little evidence that permanent closures help reef systems resist disturbances. In fact, some species of coral (particularly the branching and foliose taxa) are susceptible to climate disturbances and are also frequently damaged by fishing. Consequently, these species tend to be more abundant in closures, which can actually mean that bleaching events lead to greater coral mortality inside compared to outside closures (Graham et al. 2008b, McClanahan 2008, Darling et al. 2010).

Similar to the findings for corals, the impacts of bleaching on coral reef fish found there were actually greater losses of certain types of fishes in the closures of the WIO, specifically fish smaller than 20 cm that feed on plankton or algae (Graham et al. 2008b). An evaluation of life-history characteristics that influence extinction found that there is a trade-off among species in their susceptibility to climate and fishing disturbances (Graham et al. 2011). In short, fish species tend to be sensitive to climate disturbances or fishing, but often not both. Consequently, where fishing and climate change act together, a large variety of species may be lost. Clearly, different types of management may be required for fishes with different types of susceptibility. Species that are sensitive to fishing will benefit from closures or gear-based management. Species that are highly sensitive to climate change will thrive best in reefs with low levels of exposure to climate change. These reefs will benefit from efforts to maintain the integrity of the habitat through reduced runoff and pollution. Importantly, the species vulnerable to fishing perform critical ecological processes, such as herbivory, suggesting that even local scale efforts to manage fisheries are likely to help maintain functioning reef ecosystems (Graham et al. 2011).

There is also an emerging body of research that examines how closures affect coastal societies. The establishment of a closure invariably reallocates ownership and use rights. This reallocation can influence social, economic, and political conditions, such as employment, income, power relations, patterns of consumption, and material assets (Christie 2004, Mascia and Claus 2009). Additionally, this reallocation of use rights results in winners and losers. The winners are often those who already have power and those with high levels of adaptive capacity, who can take advantage of these changes (Christie 2004). For example, a closure in Lombok, Indonesia, marginalized fishers by restricting fishing in favor of tourism activities (Satria et al. 2006). Despite purported benefits to fisheries, the establishment of closures can actually reduce food security. For example, across 40 community-based closures in the Philippines, large fines for violations of closures rules were associated with deteriorating child nutrition status over 2 years (Gjertsen 2005). However, this is not always the case, and there are also studies that show improved nutrition after the establishment of a closure (Aswani and Furusawa 2007).

Fisheries closures can provide a powerful tool for helping to manage fisheries, with potential social and ecological benefits that can extend to the broader seascape, particularly when combined with the other management tools discussed later. However, closure is not a panacea for fisheries problems, and empirical evidence of their capacity to build resilience against climate disturbances is modest. In general, the higher the fishing intensity and the lower the standing fish biomass, the higher will be the proportion of area that is required in closures for the benefits of fisheries to accrue. In extreme circumstances, this can require that up to 35% of the fishing grounds be closed to fishing (Rodwell et al. 2002, Gell and Roberts 2003).

Given that countries such as Kenya, Reunion, and Mauritius have low levels of fish biomass and high fishing effort, up to 35% of reefs would need to be closed for this technique to enhance the recovery of fish and improve fishing in the wider seascape. This is considerably more reef area closed to fishing than any country in the region (Table 10.1). Conclusions from a mix of modeling and empirical studies would also suggest that, ideally, the size of these closures should be roughly 5 km^2 and be separated by less than 10 km. This ideal scenario is unlikely to emerge in the WIO given the social, economic, and political realities of establishing closures (Box 10.1). Even in Kenya, which has some of the most successful parks in the region, the Kenya Wildlife Service has been unable to create effective new parks since the late

TABLE 10.1

Percentage of coral reefs in Kenya, Tanzania, Madagascar, Mauritius, and the Seychelles protected by no-take fishing closures

Country	Area of coral reef (km²)[a]	No take area (NTA)(km²)[b,c]	NTA as % of reef area
Madagascar	2230	10.4	0.5%
Mauritius	870	8.5	0.9
Tanzania	3580	66	1.9
Kenya	630	54.3	8.6
Seychelles	1690	255.7	15.1

[a] (Spalding et al. 2001), [b] (Gell and Roberts 2003), [c] (Wells 2006).

Source: Adapted from McClanahan and colleagues (2008d).

BOX 10.1

Establishing a Marine Park in Mafia Island, Tanzania

The persistence of destructive dynamite fishing on some of the region's most diverse coral reefs near Mafia Island, Tanzania, led to a long process involving an international conservation organization and the Tanzanian government to establish a marine park on the southern section of this island. The park includes areas that are totally closed to fishing, areas with restricted gear use, and areas with species restrictions. The process of establishing this park has been a mix of successes and ongoing challenges for more than 15 years. The process has been well described from the perspective of marginalized communities and has been criticized for not sufficiently involving local resource users in park planning and implementation (Walley 2004). This has led to limited compliance with park management regulations. However, the communities have heterogeneous perspectives on the park, with more positive views among communities on the large island that have an urban center and substantial agricultural production and are adjacent to the gear and species-restricted area (McClanahan et al. 2009e). More negative views prevailed among communities on the small islands near the areas that had been closed to fishing (McClanahan et al. 2009e). Perceptions about the benefits of management are influenced by combinations of community experience with park management, the types of restrictions people are exposed to, and their available livelihood options.

1980s, and the last park that it successfully established (the Mombasa Marine Park) had to be scaled down after considerable conflicts with resource users. However, there are encouraging signs of this tool being utilized, such as an emerging system of small, community-based closures in Kenya.

RESTRICTIONS ON TIME

Restrictions on the times that people can fish can include formal seasonal restrictions, temporary closures, and even informal limits such as weather-related patterns (being too windy or rough to fish during certain months) and religious holidays (Cinner 2007). Cultural restrictions on when people can fish can prove very effective, as was found in the evaluation of temporary fisheries closures called taboo or *tambu* areas, which are used in Melanesia (Cinner et al. 2006, Cinner and Aswani 2007, Bartlett et al. 2009). To provide food for culturally important feasts, communities temporarily restrict fishing by placing a taboo on certain reefs (Cinner et al. 2005, Bartlett et al. 2009). The resultant harvest can be large (almost 200 kg), and ecological conditions inside these taboo reefs can be significantly improved compared to non-taboo areas (McClanahan et al. 2006, Bartlett et al. 2009). These dynamic temporal closures often find more local support among the community, which can see direct benefits from the closure; benefits from permanent closures are more indirect and difficult to appreciate (Cinner et al. 2005).

One of the key questions about temporal restrictions is whether they can provide lasting ecological benefits and contribute to the resilience of the ecosystem. Closing areas temporarily can promote the recovery of fish, but it can take a very short time to capture the gained biomass. For example, a study associated with a border change in the Mombasa Marine National Park found that the recovery of fish from many years of closure was harvested in just three months (McClanahan and Mangi 2000). A modeling approach to this problem investigated whether dynamic temporal closures can improve the abundance of herbivorous fishes (Game et al. 2009). For closure levels of 10% to 30% of the reef area, rotating protection led to herbivores being more evenly distributed across the seascape but no increases in their overall abundance. However, an increase in herbivore abundance was achieved when temporary closures were rotated among a small subset of reefs rather than throughout the whole seascape. Importantly, the social context plays a large role in whether these types of temporary closures are likely to have lasting ecological benefits. In open-access fisheries systems, fishers are more likely to intensively fish an area after temporary restrictions have been lifted. In other types of common property systems, such as where marine tenure or property rights exist, there may be more incentive for restrained harvesting (Cinner et al. 2006).

RESTRICTIONS ON SPECIES AND SIZE

Some key species are particularly vital in maintaining the important ecosystem functions in coral reefs. These include some predatory macro-invertebrate fish that feed on sea urchins and starfish, such as triggerfish and large-bodied wrasses and emperors (McClanahan 2000a, McCook et al. 2010), grazing parrotfish that play a key role in keeping the level of erect algae low (Bellwood et al. 2004), and species that may help in reversing phase shifts by eating erect macroalgae (Bellwood et al. 2007). Restricting the capture of these species is expected to maintain a coral-dominated state by avoiding or reversing the dominance of algae and sea urchins.

The application and practicality of functional group restrictions on the large scale of management have not been evaluated, but ecological modeling efforts suggest that some ecological benefits are possible (McClanahan 1995, Game et al. 2009). According to a simulation model of potential fisheries yields based on energy flows between species, not fishing the species that feed on large invertebrates will keep sea urchins numbers low and this will in turn increase the algae consumed by and the productivity of the herbivorous fishes, but it will eventually lead to a reef dominated by algae (McClanahan 1995). Not fishing herbivorous fishes to avoid this algal dominance will, however, considerably reduce the fisheries production of coral reefs because these species feed low in the food web and are fast-growing, which means they can generate a considerable portion of the fisheries yield.

Size is a critical aspect of fisheries ecology because the largest individual fishes do most of the reproduction and also play a disproportional role in key ecosystem functions (Birkeland and Dayton 2005, Lokrantz et al. 2008). For example, the body size of parrotfish has a nonlinear relationship to the amount of scraping these fish do on reefs, such that larger fish do much more scraping than smaller fish (Lokrantz et al. 2008). Protecting the largest fishes is expected to improve reproduction, recruitment, and key functions, such as herbivory.

Having minimum size limits on caught fish is one of the least contentious management options in most fisheries (McClanahan et al. 2005c, McClanahan et al. 2009e). The opposite, of having limits on the largest individuals, has, however, received less management or scientific attention in coral reef fisheries (Birkeland and Dayton 2005). In well-managed fisheries, such as the Great Barrier Reef in Australia, managers use slot limits, allowing fishers to capture intermediate-sized fish; this avoids capture of the largest individuals, which maximized their reproduction potential, and the smallest individuals, which have not reached their optimal size for harvesting or reproduction.

The difficulty with size and species limits is that coral reef fisheries are highly species diverse, typically composed of more than 100 species, and the commonly used gear, such as nets and traps, can target multiple species and sizes (McClanahan and Cinner 2008, Cinner et al. 2009a). Thus, awareness and compliance with size and species regulations would need to be high to ensure that fishers returned captured individuals with these attributes. It is through the management of specific fishing gear that the capture of certain species can generally be reduced or avoided.

RESTRICTIONS ON THE USE OF SPECIFIC FISHING GEAR

Gear-based management can be used to regulate the use of fishing gear that (1) catches the smallest or largest fish, (2) has detrimental effects on fisheries habitat, and (3) targets the fishes thought to play a critical role in the resilience of reef systems, including grazers and macro-invertivores (McClanahan and Mangi 2004, Cinner et al. 2009a). This management tool can also reduce competition for resources among fishers and result in increased catch and incomes for fishers (Box 10.2). Competition between gear types is influenced by the amount of overlap in the use of fishing grounds and the species composition of the catches, and also whether one gear type excludes the use of others (McClanahan and Mangi 2004). For example, the use of beach seine nets generally precludes the use of some other types of gear in the same area because they are actively dragged across the bottom and will disturb traps and set nets.

In a climate change context, gear-based management may help reduce climate change disturbances by limiting the mortality of species that are both susceptible to losses of coral and that promote the recovery of coral reefs (Cinner et al. 2009a). A study of artisanal fish catch in Kenya found that species highly susceptible to the effects of coral bleaching comprised less than 6%

BOX 10.2

Eliminating Beach Seine Nets in Kenya

Beach seine nets are destructive to coral reef habitat when used there, target a large proportion of juvenile fish, overlap in space and species composition, and therefore conflict with other gear and user groups (Mangi and Roberts 2006). Disagreements between local fishers, migratory fishers, and national government policies created a long struggle to remove beach seines in southern Kenya, even after they were declared illegal (McClanahan 2007). Surveys of fishers' perceptions revealed that very few people were in favor of these nets, but entrenched politics and bribery of key individuals undermined enforcement of both national law and fishermen's preferences (McClanahan et al. 2005c). The process was resolved in some fishing grounds through a series of co-management meetings where a private conservation organization, traditional and elected fisheries leaders, and the national government's fisheries department representatives agreed on the division of responsibilities. Fish being caught by seine nets were very small (about 12 cm) and many of the caught species have very fast growth potential (Kaunda-Arara and Rose 2006). Consequently, elimination of seine nets led to an increase in the catch by other gear within one year of their removal.

The gear change caused switches in the composition of the catch away from parrotfish and toward a greater diversity of species (McClanahan et al. 2008c). The combination of closures and gear restriction also increased the catch of the most valuable species and the size of the fish, which fetched higher per weight prices, and these two changes increased fisher personal incomes, particularly in fishing grounds next to the reef that had been closed to fishing (McClanahan 2010). These rapid benefits led to large-scale acceptance of the system and an eventual change that affected many of the most entrenched landing sites. The specific impact on these seine net owners and laborers was not examined, but other studies have found that seine fishers are younger, poorer, and have less capital invested in the fishery (Cinner 2010).

of the catch from any specific gear (Fig. 6.4; Cinner et al. 2009a). These small, coral-associated fish are likely to be incidental and opportunistic components of the catch rather than target species. When using traps, it may be possible to create escape gaps in the trap design so that these incidental species can escape to the water unharmed. In this same study, fish species with medium levels of susceptibility to the effects of coral bleaching represented around 40% of fishers' catch. These species are more likely to be affected by the loss of structural complexity resulting from collapse of the reef matrix than the loss of living coral (Graham et al. 2006).

There was considerable variability in the proportion of functional groups caught by the different gear (Fig. 10.3). Overall, 25% of the captured fish were reef scrapers and herbivores—fish that are considered important for the recovery of coral reef ecosystems because they reduce algae that compete with corals (Bellwood et al. 2004). Both nets and beach seines targeted a fairly diverse suite of feeding groups of fish, but there was a fairly high proportion of herbivorous fish in the catch, including grazers of erect algae and key species. Grazers were primarily targeted by spear guns, traps, and seine nets. The use of these types of fishing gear is likely to compound the effects of coral bleaching and potentially retard the ability of reefs to recover from a disturbance. In contrast, line fishing catches the lowest proportions of susceptible species and key functional groups. Thus, by this criteria line fishing may be the most preferential gear to use on reefs affected by high coral mortality.

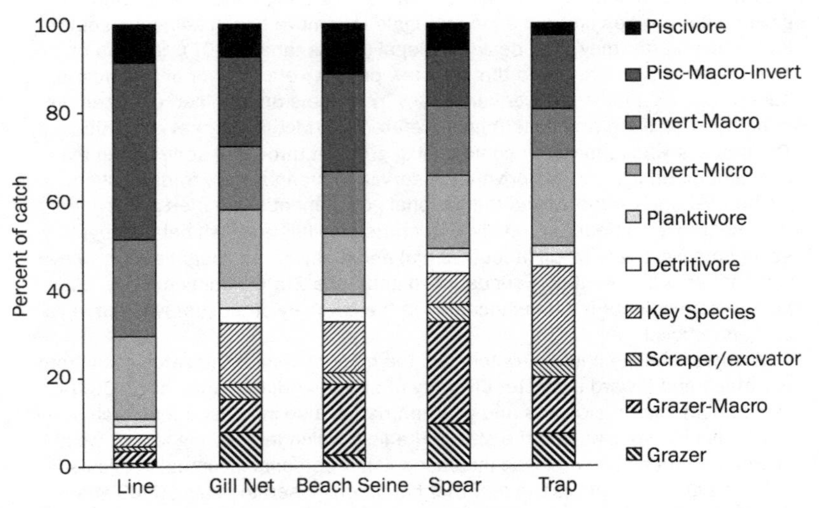

FIGURE 10.3 Catch by gear as a proportion of functional groups targeted. Types of gear are arranged from those that capture the lowest proportion of species and functional groups considered critical for recovery to disturbance on the left, to gear that capture the most on the right.

Source: Cinner et al. (2009d).

LIMITING FISHING EFFORT

Fishing effort can be restricted in a number of ways—using quotas or bag limits on catch, restricting the number of fishers (often through licensing), and developing property rights, which can be used to exclude people from accessing fishery resources. Ecological modeling and empirical fisheries studies indicate that many of the hard-to-reverse ecological changes begin in coral reefs when artisanal fisher numbers exceed four to seven fishers per km^2 of reef, which is likely to occur just below the MMSY box in Figure 10.2 (McClanahan 1995, McClanahan et al. 2008c). Consequently, by limiting fisher numbers below this level and keeping gear technology constant, it should be theoretically possible to maintain high yields and avoid unwanted ecological change.

Restricting fishing effort and the catch per person has, however, been largely unsuccessful in tropical coral reef fisheries. Controlling fishing effort through the number of licenses issued or limiting fishers' daily catch through quotas, as is frequently done in wealthy nations, have arguably been a large-scale failure in poor tropical countries. This is generally due to lack of incentives, compliance, and ineffective enforcement capabilities in many poor countries. Likewise, schemes that attempt to buy out fishers are unlikely to work in situations where there are few barriers to entering the fishery, as is the case in many parts of the WIO. Quite simply, without adequate barriers to entry, if a fisher is bought out of the fishery, someone else can readily take his place.

Managing fishing effort through the privatization of stocks has been argued by some to be one of the most successful management measures and has been used extensively in developed country fisheries (Hardin 1968, Costello et al. 2008). Privatization can take different forms, such as individually transferrable quotas, which allocate fishers a share of the yield, which they can buy or sell, or tenure over a delineated fishing ground, which allows for exclusive use. The basic premise of this management technique is that giving fishers "ownership rights" is thought to promote better stewardship, reduce overexploitation, and minimize other fishery problems and conflicts.

Some poor country fisheries have begun transitioning from open-access to property rights. For example, in the mid-1990s, the government of Chile developed delineated property rights for artisanal fisher organizations. These areas, which now cover almost 1,000 km^2, were designed mainly for the management of a benthic gastropod (*Concholepas concholepas)* but have had clear benefits to important fishery species (Gelcich et al. 2008b). There have also been successful attempts to limit access to fisheries by local communities, typically through enforcement of customary property rights. For example, in some parts of the Pacific, specific fishing grounds are owned by clans who may exclude nonmembers (Cinner and Aswani 2007). Based on traditional rights, some local groups of fishers can organize and exclude "outsiders" or ask them to pay user fees to the community (Cinner 2005, Cinner and Aswani 2007, Cinner et al. 2009f).

These allocation-based management tools can help to reduce some of the problems posed by open-access fisheries, but there are numerous cases showing that they do not promote stewardship, as is also evident from terrestrial private property, and can create social equity issues (Pitcher and Lam 2010). Privatization of fisheries may disrupt existing common property systems and shift power structures (Gelcich et al. 2006, Cinner 2007, Cinner and Aswani 2007, Cinner et al. 2007). Developing property rights systems can also create conflicts with migratory fishers and also with people outside the systems of local control (Castilla et al. 2007). Additionally, reefs under customary tenure can be heavily exploited because excluding outsiders does not necessarily mean that the users themselves will exercise self-restraint. As fishing effort has been difficult to manage, most of the progressive management systems in the region have focused on other restrictions. Nevertheless, this is among the toolbox of options that might be used depending on the local context, and it is likely that the developing Beach Management Unit (BMU) process in East Africa will eventually implement similar restrictions.

A Comparison of Management Effectiveness

Specific types of management tend to result in levels of fish biomass above, within, or below the key thresholds and MMSY window we described earlier (Fig. 10.4). There is considerable variation in how management systems operate in the different countries throughout the region, but the overall pattern is that closures have the most biomass, followed by restricted use, and then open-access fisheries (Fig. 10.4). Restricted use in this example is generally characterized by limits on the types of gear allowed but also includes restrictions on species that can be taken. For the purposes of this figure, we grouped these management tools together into the restricted use category because they rarely occur in isolation and were thus not possible to separate. On average, the restricted areas fall within the window of MMSY, indicating that these management systems can hold promise if applied successfully.

Of course, the success of this application can have considerable variation within and between countries. At the national scale, enforcement of fisheries closures in Kenya is good, but open-access fisheries have among the lowest fish biomass in the region. Some countries such as Madagascar and Mauritius exhibit few differences among management systems, suggesting low compliance with management. Some countries we surveyed such as the Maldives, Mayotte, and parts of Mozambique have light fishing on reefs and focus fishing on other systems, such as pelagic, deep-water benthic, and mangroves. Since reefs in these areas are lightly fished, differences in fish biomass between the various management systems were small.

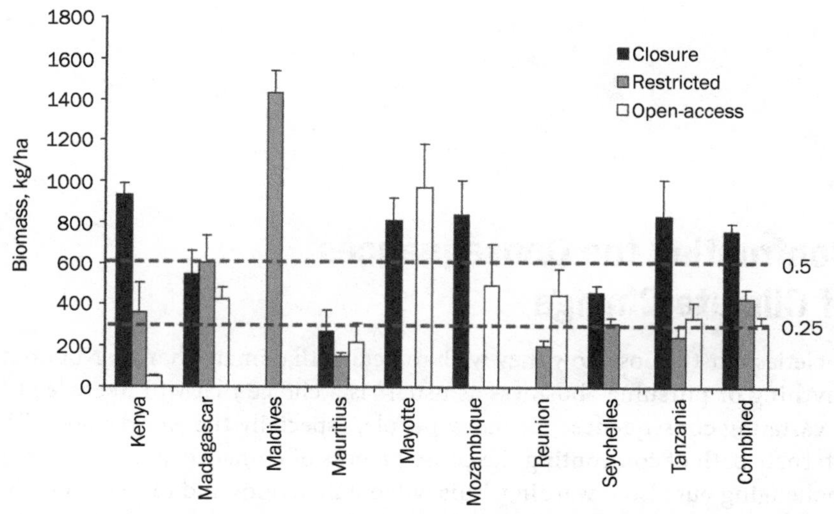

FIGURE 10.4 Fish biomass levels for open-access, fishery closures, and restricted use sites across the western Indian Ocean. The window of multi-species maximum sustained yield (MMSY) has been included to demonstrate how the different types of management may sustain fisheries.

Conclusion

The livelihoods of approximately 35 million people depend, in some part, on fisheries in the WIO region. Effectively using fisheries management tools may help to navigate climate change disturbances and will be critical to maintaining both food security and human well-being. However, regulating resource use will generally lead to periods of economic and social disturbance for some people, particularly those with the lowest levels of adaptive capacity. Consequently, initiating changes in resource use and management can be extremely difficult, particularly when there are substantial vested interests in the status quo and threats are likely to materialize over a longer time horizon, as is the case with climate change. Yet the consequences of not acting are likely to be much more severe. A trial and error process is certain to unfold as climate disturbances accelerate. Resource users, planners, and managers must analyze the context of change and use appropriate tools with high chances of success. This will ultimately require cumulative learning from both the successes and failures, but the chances of success should be improved if management uses a context-specific framework to guide efforts to build resilience in both social and ecological systems (Chapter 8).

11

Confronting the Consequences of Climate Change

Societies must choose how they wish to deal with climate change. Not doing anything or pursuing "business as usual" is a choice that will likely lead to devastating consequences for many people, especially the world's poor. The alternate path of confronting the consequences of climate change is far more challenging but also rewarding. This will entail actions and investments that are far-sighted and often difficult. Yet in the long term, the costs of not acting will be much, much higher. The debate about the causes and culpability of climate change often promoted by those with vested interests is an unfortunate distraction delaying needed actions (Billet 2009, Schneider 2009).

Confronting the consequences of climate change must occur at international, national, and local levels. International action and coordination are needed both to mitigate climate change and to coordinate the funding and efforts required for adaptation at national and local levels. Regional collaborations will help to coordinate climate-related policies, research, and training. Likewise, addressing key transboundary issues at the regional scale may have important climate adaptation implications, such as fishers crossing international borders. North-South international partnerships will also be critical for determining resource access and distribution, capacity building, and technology transfer. Transparent and properly administered and monitored international fisheries contracts can play an important role in providing adaptation funding and capacity for management-challenged regions, such as Africa (Chapter 2; Stillwell et al. 2010). These national and international actions, while critical, have been thoroughly discussed in other places and were consequently not our focus.

In this book, we highlighted a tool box of options for confronting the consequences of climate change through building local-scale adaptive capacity in societies and improving the condition of the natural resources on which people depend for their livelihoods. Building adaptive capacity will require strengthening the most appropriate and needed aspects of a society's assets, flexibility, learning, and social organizations. The ways of doing this are diverse and will, of course, depend on existing local capacities and needs. Economic development plans most frequently focus on the short-term accumulation of assets, which can often come at the price of sustaining resources, and we have therefore emphasized the three other aspects of adaptive capacity for long-term solutions. Improving the condition of resources generally requires restricting or limiting society's actions. In our example of coral reef fisheries, we described options for restricting specific fishing grounds, the time that people can fish, the gear they can use, and the sizes and species they can capture. These two broad concepts, of building social capacities and limiting certain types of resource use, interact in complicated ways that create both challenges and opportunities for adaptation.

Sometimes local situations, such as poverty traps, mean that restrictions on resource use may have disproportionate impacts on the poorest and most vulnerable, as described in Chapter 7. Other times, improvements in certain aspects of adaptive capacity can actually led to more degradation of natural resources (Cinner et al. 2009c) or responses that may amplify environmental change (Cinner et al. 2011), as described in Chapter 9. On a larger scale, fish biomass across most levels of economic development and management, apart from the Maldives and Mayotte, was below many of the important ecological thresholds highlighted in Chapter 10 (Fig. 10.4). Thus, there is often a need for coupled actions that simultaneously govern resource use and build capacity in ways that do not degrade resources. Policies and actions that focus on one without the other are common, yet avoiding poverty and environmental catastrophes is more likely with a dual focus.

One of the central themes of this book has been that adaptation solutions are context dependent, determined in part by aspects of local resource conditions, adaptive capacity, and exposure to climate change impacts, but also by people's history, culture, and aspirations. The global context in which adaptation occurs is also critical. There is a need to consider these national and international issues because, as we saw in Chapter 2, this broader context can greatly influence the direction of proposed management actions as well as develop a funding vehicle for local management. A key part of the context-dependent nature of adaptation is that solutions developed in one location should not necessarily be seen as a blueprint to be indiscriminately applied to other places. The planet is littered with these blueprint failures.

This complicated context-dependent nature of adaptations should not, however, be seen as justification for cynicism and inaction. Applying the framework we develop in this book can help to provide governments, scientists, managers, and donors with critical information about the local context and develop nuanced actions that reflect these local conditions. This information can help to identify key opportunities and narrow the range of potential adaptation options that may be suitable for a particular location. An example of this type of opportunity is the capacity for more fisheries closures in the countries of Mauritius and Reunion. The high adaptive capacity and low national reliance on coral reef fisheries can help protect these regionally important reefs and possibly stimulate these countries as ecotourism destinations where low climate disturbances are predicted. Alternatively in places such as Kenya, where the exposure to bleaching is much higher and adaptive capacity is lower, existing protected areas will need to be complemented by management options that will enhance the condition of the broader seascape but have fewer impacts on resource users. Sites that have effectively removed destructive seine nets have shown increased catches and incomes for fishers. The benefits from this type of management could extend through the country at low costs while simultaneously building aspects of fishers' adaptive capacity. In the examples given, investments in social organization, learning, and flexibility will be required to strengthen the institutions and create conditions for effective management.

Of course, it is not scientists, donors, and governments alone that influence adaptation decisions. Incorporating the ideas and aspirations of people who use local resources must be central to any adaptation strategy. Decades of experience in international development have shown that local involvement does not guarantee the success of development projects, but ignoring people's ideas and capacities regularly leads to failure. On the other hand, although local involvement and context are critical, they are likely insufficient on their own for successful resource governance or climate change adaptation. Policies and programs at the local level must be supported on national and international scales. For example, local attempts to manage resources must be incorporated into a larger governance framework that supports these local actions by providing legitimacy, legal authority, and enforcement capacity. Where these larger scale linkages are missing, local actions can often be futile.

Building adaptive capacity and rebuilding depleted natural resources can take decades. Success may require incremental action, which may not necessarily occur in the order that donors, scientists, and managers would prefer. For example, donors and NGOs may wish to establish a network of protected areas to increase the resilience of marine ecosystems to the effects of climate change. However, building aspects of learning and social organization that create institutions capable of effectively restricting resource use and awareness about climate change may be required before protected areas can be established. Protecting areas without consultation and before these key capacities are developed may lead to poor compliance, which undermines both the process and outcome.

We used coral reefs as a focal lens through which to view the issues surrounding climate change and society because it is a system we know well and one of the more sensitive ecosystems, which can provide a warning sign for future changes to other social-ecological systems. The effects of climate change on other social-ecological linkages, such as agriculture, will likely have larger implications for society than our coral reef example. Nevertheless, the framework we developed in this book to explore the multidisciplinary aspects of climate change is generic in the sense that it does not necessarily apply just to coral reef ecosystems. This framework can be used to explore different social-ecological systems and types of environmental change at a range of scales. Our hope is that future applications will help societies better confront the consequences of climate change.

BIBLIOGRAPHY

Aaheim, A., and L. Sygna. 2000. Economic impacts of climate change on tuna fisheries in Fiji Islands and Kiribati. Report 2000: 4, CICERO, Oslo, Norway.

Abram, N. J., M. K. Gagan, J. E. Cole, W. S. Hantoro, and M. Mudelsee. 2008. Recent intensification of tropical climate variability in the Indian Ocean. Nature Geoscience **1**:849–853.

Adato, M., M. R. Carter, and J. May. 2006. Exploring poverty traps and social exclusion in South Africa using qualitative and quantitative data. Journal of Development Studies **42**:226–247.

Adger, N. W. 2006. Vulnerability. Global Environmental Change **16**:268–281.

Adger, W.N. 1999. Social vulnerability to climate change and extremes in coastal Vietnam. World Development **2**:249–269.

Adger, W.N. 2000. Social and ecological resilience: are they related? Progress in Human Geography **24**:347–364.

Adger, W.N. 2003. Social capital, collective action and adaptation to climate change. Economic Geography **79**:387–404.

Adger, W.N., K. Brown, and E. Tompkins. 2005b. The political economy of cross-scale networks in resource co-management. Ecology & Society **10**:9.

Adger, W.N., and M. Kelly. 1999. Social vulnerability to climate change and architecture of entitlements. Mitigation and Adaptation Strategies for Global Change **4**:253–266.

Adger, W.N., and K. Vincent. 2005. Uncertainty in adaptive capacity. Comptes Rendus Geoscience **337**:399–410.

Adger, W. N., N. W. Arnell, and E. L. Tompkins. 2005a. Successful adaptation to climate change across scales. Global Environmental Change **15**:77–86.

Adger, W. N., T. P. Hughes, C. Folke, S. R. Carpenter, and J. Rockstrom. 2005c. Social-ecological resilience to coastal disasters. Science **309**:1036–1039.

Agnew, D. J., J. Pearce, G. Pramod, T. Peatman, R. Watson, J. R. Beddington, and T. J. Pitcher. 2009. Estimating the worldwide extent of illegal fishing. PLoS One **4**:e4570.

Agrawal, A. 2001. Common property institutions and sustainable governance of resources. World Development **29**:1649–1672.

Alder, J., S. Cullis-Suzuki, V. Karpouzi, K. Kaschner, S. Mondoux, W. Swartz, P. Trujillo, R. Watson, and D. Pauly. 2010. Aggregate performance in managing marine ecosystems of 53 maritime countries. Marine Policy **34**:468–476.

Alder, J., and R. U. Sumaila. 2004. Western Africa: A fish basket of Europe past and present. Journal of Environment & Development **13**:156–178.

Allan, R. P., and B. J. Soden. 2008. Atmospheric warming and the amplification of precipitation extremes. Science **321**:1481–1484.

Allen, G. R. 2008. Conservation hotspots of biodiversity and endemism for Indo-Pacific coral reef fishes. Aquatic Conservation: Marine and Freshwater Ecosystems **18**:541–556.

Alley, R. B. 2000. Ice-core evidence of abrupt climate changes. Proceedings of the National Academy of Sciences of the United States of America **97**:1331–1334.

Allison, E., W. Adger, M.-C. Badjeck, K. Brown, D. Conway, N. Dulvy, A. Halls, A. Perry, and J. Reynolds. 2005. Effects of climate change on the sustainability of capture and enhancement fisheries important to the poor: analysis of the vulnerability and adaptability of fisherfolk living in poverty. Summary Report R4778J, Fisheries Management Science Programme/Department for International Development.

Allison, E., N. Andrew, and J. Oliver. 2007. Enhancing the resilience of inland fisheries and aquaculture systems to climate change. SAT ejournal **4**:1–35.

Allison, E., and F. Ellis. 2001. The livelihoods approach and management of small-scale fisheries. Marine Policy **25**:377–388.

Allison, E., A. Perry, M.-C. Badjeck, W. Adger, K. Brown, D. Conway, A. Halls, G. Pilling, J. Reynolds, N. Andrew, and N. Dulvy. 2009. Vulnerability of national economies to the impact of climate change on fisheries. Fish and Fisheries **10**:173–196.

Anderson, D. M., J. T. Overpeck, and A. K. Gupta. 2002. Increase in the Asian Southwest Monsoon during the past four centuries. Science **297**:596–599.

Armitage, D., R. Plummer, F. Berkes, R. Arthur, A. T. Charles, I. Davidson-Hunt, A.P. Diduck, N.C. Doubleday, D.S. Johnson, M. Marschke, P. McConney, E. Pinkerton, and E. Wollenberg. 2009. Adaptive co-management for social–ecological complexity. Frontiers in Ecology and the Environment **7**:95–102.

Armstrong, C. W., U. R. Sumaila, A. Erastus, and O. Msiska. 2004. Benefits and costs of the Namibianisation policy. Pages 203–214 *in* U. R. Sumaila, D. Boyer, M. D. Skogen, and S. I. Steinshamn, editors. Namibia's Fisheries: Ecological, Economic and Social Aspects. Eburon, Amsterdam.

Arnell, N. W. 2006. Climate change and water resources: a global perspective. Pages 167–175 *in* H. J. Schellnhuber, W. Cramer, N. Nakicenovic, T. Wigley, and G. Yohe, editors. Avoiding Dangerous Climate Change. Cambridge University Press, Cambridge.

Arrow, K., B. Bolin, R. Costanza, P. Dasgupta, C. Folke, C. Hotelling, B.-O. Jansson, S. Levin, K.-G. Maler, C. Perrings, and D. Pimentel. 1995. Economic growth, carrying capacity and the environment. Science **268**:520–521.

Asche, F., and M. Smith. 2010. Trade and Fisheries: Key Issues for the World Trade Organization. World Trade Organization, Geneva.

Aswani, S., and T. Furusawa. 2007. Do MPAs affect human health and nutrition? A comparison among villages in Roviana, Solomon Islands. Coastal Management **35**:545–565.

Aswani, S., and R. Hamilton. 2004. Integrating indigenous ecological knowledge and customary sea tenure with marine and social science for conservation of bumphead parrotfish (*Bolbometopon muricatum*) in the Roviana lagoon, Solomon Islands. Environmental Conservation **31**:69–83.

Ateweberhan, M., and T. R. McClanahan. 2010. Historical sea-surface temperature variability predicts climate change-induced coral mortality. Marine Pollution Bulletin **60**:964–970.

Ayre, D. J., and T. P. Hughes. 2004. Climate change, genotypic diversity and gene flow in reef-building corals. Ecology Letters **7**:273–278.

Babcock, R., A. Alcala, N. Barrett, G. Edgar, K. Lafferty, T. McClanahan, G. Russ, and N. Shears. 2010. Decadal trends in marine reserves: differential rates of change for direct and indirect effects. Proceedings of the National Academy of Science **107**:18256-18261.

Badjeck, M.-C., E. H. Allison, A. S. Halls, and N. K. Dulvy. 2010. Impacts of climate variability and change on fishery-based livelihoods. Marine Policy **34**:375–383.

Baettig, M., M. Wild, and D. Imboden. 2007. A climate change index: where climate change may be most prominent in the 21st century. Geophysical Research Letters **34**:L01705.

Baird, A. H., R. Bhagooli, P. J. Ralph, and S. Takahashi. 2009. Coral bleaching: the role of the host. Trends in Ecology and Evolution **24**:16–20.

Baird, A. H., and P. A. Marshall. 2002. Mortality, growth and reproduction in scleractinian corals following bleaching on the Great Barrier Reef. Marine Ecology Progress Series **237**:133–141.

Baker, A. C., P. W. Glynn, and B. Riegl. 2008. Climate change and coral reef bleaching: An ecological assessment of long-term impacts, recovery trends and future outlook. Estuarine, Coastal and Shelf Science **80**:435–471.

Baker, A. C., and A. M. Romanski. 2007. Multiple symbiotic partnerships are common in scleractinian corals, but not in octocorals: Comment on Goulet (2006). Marine Ecology Progress Series **335**:237–242.

Barange, M., and R. I. Perry. 2009 Physical and ecological impacts of climate change relevant to marine and inland capture fisheries and aquaculture. Pages 7–106 *in* K. Cochrane, C. De Young, D. Soto, and T. Bahri, editors. Climate change implications for fisheries and aquaculture: overview of current scientific knowledge. FAO Fisheries and Aquaculture Technical Paper. No. 530, Rome.

Bard, E., F. Rostek, and C. Sonzogni. 1997. Interhemispheric synchrony of the last deglaciation inferred from alkenone palaeothermometry. Nature **385**: 707–710.

Barnett, J. 2001. Security and climate change. Working Paper 7, Tyndall Centre, Norwich, UK.

Barnett, J., and W. N. Adger. 2007. Climate change, human security and violent conflict. Political Geography **26**:639–655.

Barrett, C., and M. Carter. 2001. Can't get ahead for falling behind: new directions for development policy to escape poverty and relief traps. Choices **17**:35–38.

Barrett, C., M. Carter, and P. Little. 2006a. Understanding and reducing persistent poverty in Africa: Introduction to a special issue. Journal of Development Studies **42**:167–177.

Barrett, C. B., P. P. Marenya, J. G. McPeak, B. Minten, F. M. Murithi, W. Oluoch-Kosura, F. Place, J. C. Randrianarisoa, J. Rasambainarivo, and J. Wangila. 2006b. Welfare dynamics in rural Kenya and Madagascar. Journal of Development Studies **42**:248–277.

Bartlett, C. Y., C. Manua, J. Cinner, S. Sutton, R. Jimmy, R. South, J. Nilsson, and J. Raina. 2009. Comparison of outcomes of permanently closed and periodically harvested coral reef reserves. Conservation Biology **23**:1475–1484.

Baschieri, A., and S. Kovats. 2010. Climate and child health in rural areas of low and middle income countries: A review of epidemiological evidence. International Public Health Journal **2**: 431–445.

Baskett, M. L., R. M. Nisbet, C. V. Kappel, P. J. Mumby, and S. D. Gaines. 2010. Conservation management approaches to protecting the capacity of corals to respond to climate change: a theoretical comparison. Global Change Biology **16**:1229–1246.

Becker, C., and E. Ostrom. 1995. Human ecology and resource sustainability: The importance of institutional diversity. Annual Review of Ecological Systems **26**:113–133.

Bellwood, D. R., R. H. Hughes, and A. S. Hoey. 2007. Sleeping functional group drives coral-reef recovery. Current Biology **16**:2434–2439.

Bellwood, D. R., T. P. Hughes, C. Folke, and M. Nystrom. 2004. Confronting the coral reef crisis. Nature **429**:827–832.

Bender, M. A., T. R. Knutson, R. E. Tuleya, J. J. Sirutis, G. A. Vecchi, S. T. Garner, and I. M. Held. 2010. Modeled impact of anthropogenic warming on the frequency of intense Atlantic hurricanes. Science **327**:454–458.

Berkelmans, R., G. De'ath, S. Kininmonth, and W. J. Skirving. 2004. A comparison of the 1998 and 2002 coral bleaching events on the Great Barrier Reef: spatial correlation, patterns, and predictions. Coral Reefs **23**:74–83.

Berkes, F., T. P. Hughes, R. S. Steneck, J. A. Wilson, D. R. Bellwood, B. Crona, C. Folke, L. H. Gunderson, H. M. Leslie, J. Norberg, M. Nystrom, P. Olsson, H. Osterblom, M. Scheffer, and B. Worm. 2006. Globalization, roving bandits, and marine resources. Science **311**:1557–1558.

Berkes, F., and C. Seixas. 2006. Building resilience in lagoon social-ecological systems: a local-level perspective. Ecosystems **8**:967–974.

Bernstein, H., B. Crow, and H. Johnson. 1992. Rural Livelihoods: Crises and Responses. Oxford University Press and the Open University, Oxford, UK.

Berumen, M. L., and M. S. Pratchett. 2006. Recovery without resilience: Persistent disturbance and long-term shifts in the structure of fish and coral communities at Tiahura Reef, Moorea. Coral Reefs **25**:647–653.

Billett, S. 2009. Dividing climate change: global warming in the Indian mass media. Climate Change **99**:295–329.

Birkeland, C., and P. Dayton. 2005. The importance in fishery management of leaving the big ones. Trends in Ecology and Evolution **20**:356–358.

Bodin, O., O. Teng, A. Norman, J. Lundberg, and T. Elmqvis. 2006. The value of small size? loss of forest patches and ecological thresholds in southern Madagascar. Ecological Applications **16**:440–451.

Boko, M., I. Niang, A. Nyong, C. Vogel, A. Githeko, M. Medany, B. Osman-Elasha, R. Tabo, and P. Yanda. 2007. Africa. Pages 433–467 *in* M. Parry, O. Canziani, J. Palutiko, P. van der Linden, and C. Hanson, editors. Climate Change 2007: Impacts, Adaptation and Vulnerability. Contribution of Working Group II to the Fourth Assessment Report of the Intergovernmental Panel on Climate Change. Cambridge University Press, Cambridge, UK,.

Bowden, J. H., and F. H. M. Semazzi. 2007. Empirical analysis of intraseasonal climate variability over the Greater Horn of Africa. Journal of Climate **20**:5715–5731.

Brahmananda Rao, V., C. C. Ferreira, S. H. Franchito, and S. S. V. S. Ramakrishna. 2008. In a changing climate weakening tropical easterly jet induces more violent tropical storms over the north Indian Ocean. Geophysical Research Letters **35**:4.

Brewer, T., J. Cinner, A. Green, and J. Pandolfi. 2009. Thresholds and multiple scale interaction of environment, resource use, and market proximity on reef fishery resources in the Solomon Islands. Biological Conservation **142**:1797–1807.

Brooks, N., W. Adger, and M. Kelly. 2005. The determinants of vulnerability and adaptive capacity at the national level and the implications for adaptation. Global Environmental Change **15**:151–163.

Brown, M. E., and C. C. Funk. 2008. Food security under climate change. Science **319**:580–581.

Browne, A., and H. Barrett. 1991. Female education in Sub-Saharan Africa: the key to development? Comparative Education **27**:275–285.

Bruno, J. F., and E. R. Selig. 2007. Regional decline of coral cover in the Indo-Pacific: Timing, extent, and subregional comparisons. PLoS One **2**:e711.

Bruno, J. F., H. Sweatman, W. F. Precht, E. R. Selig, and G. W. Schutte. 2009. Assessing evidence of phase shifts from coral to macroalgal dominance on coral reefs. Ecology **90**:1478–1484.

Bryceson, D. 2002. The scramble in Africa: reorienting rural livelihoods. World Development **30**:725–739.

Burkepile, D. E., and M. E. Hay. 2006. Herbivore vs. nutrient control of marine primary producers: context-dependent effects. Ecology **87**:3128–3139.

Caner, L., D. L. Seen, Y. Gunnell, B. R. Ramesh, and G. Bourgeon. 2007. Spatial heterogeneity of land cover response to climatic change in the Nilgiri highlands (Southern India) since the last glacial maximum. The Holocene **17**:195–205.

Cantin, N. E., A. L. Cohen, K. B. Karnauskas, A. M. Tarrant, and D. C. McCorkle. 2010. Ocean warming slows coral growth in the Central Red Sea. Science **329**:322–325.

Carilli, J. E., R. D. Norris, B. Black, S. M. Walsh, and M. McField. 2010. Century-scale records of coral growth rates indicate that local stressors reduce coral thermal tolerance threshold. Global Change Biology **16**:1247–1257.

Carpenter, R. C. 1990. Mass mortality of *Diadema antillarum*: II effects on population densities and grazing intensities of parrotfishes and surgeonfishes. Marine Biology **104**:79–86.

Carpenter, S., B. Walker, M. Anderies, and N. Abel. 2001. From metaphor to measurement: resilience of what to what? Ecosystems **4**:765–781.

Carreiro-Silva, M., and T. R. McClanahan. 2001. Echinoid bioerosion and herbivory on Kenyan coral reefs: The role of protection from fishing. Journal of Experimental Marine Biology and Ecology **262**:133–153.

Carricart-Ganivet, J. P. 2007. Annual density banding in massive coral skeletons: result of growth strategies to inhabit reefs with high microborers' activity? Marine Biology **153**:1–5.

Carter, M., and C. Barrett. 2006. The economics of poverty traps and persistent poverty: an asset-based approach. Journal of Development Studies **42**:178–199.

Carter, M., and J. May. 2001. One kind of freedom: poverty dynamics in post-apartheid South Africa. World Development **29**:1987–2006.

Castañeda, I. S., J. P. Werne, and T. C. Johnson. 2007. Wet and arid phases in the southeast African tropics since the Last Glacial Maximum. Geology **35**:823–826.

Castilla, J. C., S. Gelcich, and O. Defeo. 2007. Successes, lessons, and projections from experience in marine benthic invertebrate artisanal fisheries in Chile. Pages 25–42 *in* T. McClanahan and J. C. Castilla, editors. Fisheries Management: Progress towards Sustainability. Blackwell, Oxford, UK.

Cesar, H., L. Burke, and L. Pet-Soede. 2003. The economics of worldwide coral reef degradation. Cesar Environmental Economics Consulting, Arnhem.

Charles, A. T. 1992. Fishery conflicts: a unified framework. Marine Policy **16**:379–393.

Cheal, A. J., S. K. Wilson, M. J. Emslie, A. M. Dolman, and H. Sweatman. 2008. Responses of reef fish communities to coral declines on the Great Barrier Reef. Marine Ecology Progress Series **372**:211–223.

Cheung, W., V. Lam, J. Sarmiento, K. Kearney, R. Watson, D. Zeller, and D. Pauly. 2010. Large-scale redistribution of maximum fisheries catch potential in the global ocean under climate change. Global Change Biology **16**:24–35.

Christie, P. 2004. MPAs as biological successes and social failures in Southeast Asia. Page 275 *in* J. B. Shipley, editor. Aquatic Protected Areas as Fisheries Management Tools: Design, Use, and Evaluation of These Fully Protected Areas. American Fisheries Society, Bethesda.

Cinner, J. 2005. Socioeconomic factors influencing customary marine tenure in the Indo-Pacific. Ecology and Society **10**:1–14

Cinner, J. 2007. The role of taboos in conserving coastal resources in Madagascar. Traditional Marine Resource Management and Knowledge Information Bulletin **22**:15–23.

Cinner, J. 2009. Migration and coastal resource use in Papua New Guinea. Ocean & Coastal Management **52**:411–416.

Cinner, J. 2010. Poverty and the use of destructive fishing gear near east African marine protected areas. Environmental Conservation **4**: 321–326.

Cinner, J., and S. Aswani. 2007. Integrating customary management into marine conservation. Biological Conservation **140:** 201–216.

Cinner, J., and O. Bodin. 2010. Livelihood diversification in tropical coastal communities: a network-based approach to analyzing "livelihood landscapes." PLoS One **5**:e11999.

Cinner, J., T. Daw, C. Folke, and C. Hicks. 2011. Responding to change: using scenarios to understand how socioeconomic factors may influence amplifying or dampening exploitation feedbacks among Tanzanian fishers. Global Environmental Change. **21**:7–12.

Cinner, J., T. Daw, and T. McClanahan. 2009e. Socioeconomic factors that affect artisanal fishers' readiness to exit a declining fishery. Conservation Biology **23**:124–130.

Cinner, J., M. Fuentes, and H. Randriamahazo. 2009b. Exploring social resilience in Madagascar's marine protected areas. Ecology & Society **14**:41.

Cinner, J. E., M. J. Marnane, and T. R. McClanahan. 2005. Conservation and community benefits from traditional coral reef management at Ahus Island, Papua New Guinea. Conservation Biology **19**:1714–1723.

Cinner, J., M. Marnane, T. McClanahan, and G. Almany. 2006. Periodic closures as adaptive coral reef management in the Indo-Pacific. Ecology & Society **11**:31.

Cinner, J., and T. McClanahan. 2006. Socioeconomic factors that lead to overfishing in small-scale coral reef fisheries of Papua New Guinea. Environmental Conservation **33**:73–80.

Cinner, J., T. McClanahan, T. Daw, N. Graham, J. Maina, S. Wilson, and T. Hughes. 2009c. Linking social and ecological systems to sustain coral reef fisheries. Current Biology **19**:206–212.

Cinner, J., T. McClanahan, N. Graham, M. Pratchett, S. Wilson, and J. Raina. 2009a. Gear-based fisheries management as a potential adaptive response to climate change and coral mortality. Journal of Applied Ecology **46**:724–732.

Cinner, J., and R. Pollnac. 2004. Poverty, perceptions and planning: why socioeconomics matter in the management of Mexican reefs. Ocean & Coastal Management **47**:479–493.

Cinner, J. E., S. G. Sutton, and T. G. Bond. 2007. Socioeconomic thresholds that affect use of customary fisheries management tools. Conservation Biology **21**:1603–1611.

Cinner, J., A. Wamukota, H. Randriamahazo, and A. Rabearisoa. 2009d. Toward community-based management of inshore marine resources un the Western Indian Ocean. Marine Policy **33**:489–496.

Clark, D., S. Brown, D. Kicklighter, J. Chambers, J. Thomlinson, J. Ni, and E. Holland. 2001. Net primary production in tropical forests: an evaluation and synthesis of existing field data. Ecological Applications **11**:371–384.

Claudet, J., C. W. Osenberg, L. Benedetti-Cecchi, P. Domenci, J.-A. Garcia-Charton, A. Perez-Ruzafa, F. Badalamenti, J. Bayle-Sempere, A. Brito, F. Bulleri, J.-M. Culioli, M. Dimech, J. M. Falcon, I. Guala, M. Milazzo, J. Sanchez-Meca, P. J. Somerfield, B. Stobart, F. Vanderperre, C. Valle, and S. Planes. 2008. Marine reserves: size and age do matter. Ecology Letters **11**:481–489.

Clausen, R., and R. York. 2008a. Economic growth and marine biodiversity: influence of human social structure on decline of marine trophic levels. Conservation Biology **22**:458–466.

Clausen, R., and R. York. 2008b. Global biodiversity decline of marine and freshwater fish: a cross-national analysis of economic, demographic, and ecological influences. Social Science Research **37**:1310–1320.

Cohen, A. L., and M. Holcomb. 2009. Why corals care about ocean acidification: uncovering mechanisms. Oceanography **22**:118–127.

Coles, S. L., and B. E. Brown. 2003. Coral bleaching—capacity for acclimatization and adaptation. Advances in Marine Biology **46**:183–223.

Connell, J. H. 1997. Disturbance and recovery of coral assemblages. Coral Reefs **16**:S101–S113.

Cook, E. R., K. J. Anchukaitis, B. M. Buckley, R. D. D'Arrigo, G. C. Jacoby, and W. E. Wright. 2010. Asian monsoon failure and megadrought during the last millennium. Science **328**:486–489.

Cooper, T. F., G. De'ath, K. E. Fabricius, and J. M. Lough. 2008. Declining coral calcification in massive Porites in two nearshore regions of the northern Great Barrier Reef. Global Change Biology **14**:529–538.

Correge, T., M. K. Gagan, J. W. Beck, G. S. Burr, G. Cabioch, and F. Le Cornec. 2004. Interdecadal variation in the extent of South Pacific tropical waters during the Younger Dryas event. Nature **428**:927–929.

Costello, C., S. D. Gaines, and J. Lynham. 2008. Can catch shares prevent fisheries collapse? Science **321**:1678–1681.

Coughanowr, C., M. Ngoile, and O. Linden. 1995. Coastal zone management in Eastern Africa including the island states: a review of issues and initiatives. Ambio **24**:448–457.

Coulthard, S. 2008. Adapting to environmental change in artisanal fisheries—insights from a South Indian lagoon. Global Environmental Change **18**:479–489.

Crona, B., M. Nyström, C. Folke, and N. Jiddawi. 2010. Middlemen, a critical social-ecological link in coastal communities of Kenya and Zanzibar. Marine Policy **34**: 761–771.

Crueger, T., E. Roeckner, T. Raddatz, T. Schnur, and P. Wetzel. 2008. Ocean dynamics determine the response of oceanic CO_2 uptake to climate change. Climate Dynamics **31**:151–168.

Curran, S., and T. Agardy. 2002. Common property systems, migration, and coastal ecosystems. Ambio **31**:303–305.

Damassa, T. D., J. E. Cole, H. R. Barnett, T. R. Ault, and T. R. McClanahan. 2006. Enhanced multidecadal climate variability in the seventeeth century from coral isotope records in the Western Indian Ocean. Paleoceanography **21**:PA 2016.

Darling, E. S., and I. M. Cote. 2008. Quantifying the evidence for ecological synergies. Ecology Letters **11**:1278–1286.

Darling, E. S., T. R. McClanahan, and I. M. Cote. 2009. Antagonistic interaction between bleaching and fishing on coral communities. Conservation Letters **3**: 122–130.

Dasgupta, P. 1997. Nutritional status, the capacity for work, and poverty traps. Journal of Econometrics **77**:5–37.

Davies, S. 1996. Adaptive Livelihoods. Macmillan, London.

Daw,T., W.N. Adger, K. Brown, and M. Badjeck. 2009. Climate change and capture fisheries. Climate change implications for fisheries and aquaculture. Overview of current scientific knowledge., Food and Agricultural Organization, Rome.

De'ath, G., J. M. Lough, and K. E. Fabricius. 2009. Declining coral calcification on the Great Barrier Reef. Science **323**:116–119.

Dearing, J., R. Jones, J. Shen, X. Yang, J. Boyle, G. Foster, D. Crook, and M. Elvin. 2008. Using multiple archives to understand past and present climate-human-environment interactions: the lake Erhai catchment, Yunnan Province, China. Journal of Paleolimnology **40**:3–31.

Diamond, J. 2005. Collapse: how societies choose to fail or survive. Penguin, London, UK.

Diaz-Pulido, G., L. J. McCook, S. Dove, R. Berkelmans, G. Roff, D. I. Kline, S. Weeks, R. D. Evans, D. H. Williamson, and O. Hoegh-Guldberg. 2009. Doom and boom on a resilient reef: climate change, algal overgrowth and coral recovery. PLoS One **4**:e5239.

Dixson, D. L., P. L. Munday, and G. P. Jones. 2010. Ocean acidification disrupts the innate ability of fish to detect predator olfactory cues. Ecology Letters **13**:68–75.

Donner, S. D., T. R. Knutson, and M. Oppenheimer. 2007. Model-based assessment of the role of human-induced climate change in the 2005 Caribbean coral bleaching event. Proceedings of the National Academy of Sciences **104**:5483–5488.

Donner, S. D., W. J. Skirving, C. M. Little, M. Oppenheimer, and O. Hoegh-Guldberg. 2005. Global assessment of coral bleaching and required rates of adaptation under climate change. Global Change Biology **11**:2251–2265.

Dore, J. E., R. Lukas, D. W. Sadler, M. J. Church, and D. M. Karl. 2009. Physical and biogeochemical modulation of ocean acidification in the central North Pacific. Proceedings of the National Academy of Sciences **106**:12235–12240.

Duffy, R. 2006. Global environmental governance and the poitics of ecotourism in Madagascar. Journal of Ecotourism **5**:128–144.

Dulvy, N., R. Freckleton, and N. Polunin. 2004. Coral reef cascades and the indirect effects of predator removal by exploitation. Ecology Letters **7**:410–416.

Dulvy, N. K., Y. Sadovy, and J. D. Reynolds. 2003. Extinction vulnerability in marine populations. Fish and Fisheries **4**:25–64.

Eakin, C. M., J. M. Lough, and S. F. Heron. 2009. Climate variability and change: monitoring data and evidence for increased coral bleaching stress. Pages 41–67 *in* M. van Oppen and J. M. Lough, editors. Coral Bleaching: Patterns, Processes, Causes and Consequences. Springer Ecological Studies, Berlin.

Easterling, W. E., P. K. Aggarwal, P. Batima, K. M. Brander, L. Erda, S. M. Howden, A. Kirilenko, J. Morton, J.-F. Soussana, J. Schmidhuber, and F. N. Tubiello. 2007. Food, fibre and forest products. Pages 273–313 *in* M. L. Parry, O. F. Canziani, J. P. Palutikof, P. J. van der Linden, and C. E. Hanson, editors. Climate Change 2007: Impacts, Adaptation and Vulnerability: Contribution of Working Group II to the Fourth Assessment Report of the Intergovernmental Panel on Climate Change. Cambridge University Press, Cambridge.

Edmunds, P. J., and R. D. Gates. 2008. Acclimatization in tropical reefs corals. Marine Ecology Progress Series **361**:307–310.

Ehrlich, P., and A. Ehrlich. 2004. One with Nineveh—Politics, Consumption, and the Human Future. Island Press, Washington, DC.

Elasha, B., M. Medany, I. Niang-Diop, T. Nyong, R. Tabo, and C. Vogel. 2006. Impacts, vulnerability and adaptation to climate change in Africa. Background Paper for the African Workshop on Adaptation Implementation of Decision 1/CP.10 of the UNFCCC Convention, UNFCCC.

Elmqvist, T. 2004. The forgotten dry forest of southern Madagascar. Plant Talk **35**:29–31.

Emslie, M. J., A. J. Cheal, H. Sweatman, and S. Delean. 2008. Recovery from disturbance of coral and reef fish communities on the Great Barrier Reef, Australia. Marine Ecology Progress Series **371**:177–190.

EU. 2009. Fishery Statistics. European Union. http://epp.eurostat.ec.europa.eu/statistics_explained/index.php/Fishery_statistics

Fabricius, K. E. 2005. Effects of terrestrial runoff on the ecology of corals and coral reefs: review and synthesis. Marine Pollution Bulletin **50**:125–146.

Fagoonee, I., and D. Daby. 1995. Coastal zone management in Mauritius. Coastal Management Center, Manila.

FAO. 2007. Information and communication technologies benefit fishing communities. New directions in fisheries—a series of policy briefs on development issues 09, Food and Agriculture Organisation, Rome.

FAO. 2008. FISHSTAT Plus: Universal software for fishery statistical time series. FAO Fisheries Department, Fishery Information, Data and Statistics Unit 2004. Food and Agriculture Organisation, Rome.

FAO. 2009. FAO yearbook. Fishery and Aquaculture Statistics, 2007. FAO Fisheries and Aquaculture Information and Statistics Service, Rome, Italy.

Fazey, I., J. Fazey, J. Fischer, K. Sherren, J. Warren, R. Noss, and S. Dovers. 2007. Adaptive capacity and learning to learn as leverage for social-ecological resilience. Frontiers in Ecology and the Environment **5**:375–380.

Feary, D. A. 2007. The influence of resource specialization on the response of reef fish to coral disturbance. Marine Biology **153**:153–161.

Fisher, J. A. D., K. T. Frank, and W. C. Leggett. 2010. Global variation in marine fish body size and its role in biodiversity-ecosystem functioning. Marine Ecology Progress Series **405**:1–13.

Fleitmann, D., S. J. Burns, A. Mangini, M. Mudelsee, J. Kramers, I. Villa, U. Neff, A. A. Al-Subbary, A. Buettner, D. Hippler, and A. Matter. 2007. Holocene ITCZ and Indian monsoon dynamics recorded in stalagmites from Oman and Yemen (Socotra). Quaternary Science Reviews **26**:170–188.

Fleitmann, D., S. J. Burns, M. Mudelsee, U. Neff, J. Kramers, A. Mangini, and A. Matter. 2003. Holocene forcing of the Indian Monsoon recorded in a stalagmite from Southern Oman. Science **300**:1737–1739.

Folke, C., S. Carpenter, T. Elmqvist, L. Gunderson, C. S. Holling, and B. Walker. 2002. Resilience and sustainable development: Building adaptive capacity in a world of transformations. Ambio **31**:437–440.

Folke, C., T. Hahn, P. Olsson, and J. Norberg. 2005. Adaptive governance of social-ecological systems. Annual Review of Environment and Resources **30**:441–473.

Froese, R., and A. Proelß. 2010. Rebuilding fish stocks no later than 2015: will Europe meet the deadline? Fish and Fisheries **11**:194–202.

Fulanda, B., C. Munga, J. Ohtomi, M. Osore, J. Mugo, and Y. Hossain. 2009. The structure and evolution of the coastal migrant fishery of Kenya. Ocean and Coastal Management **52**:459–466.

Funk, C., M. D. Dettinger, J. C. Michaelsen, J. P. Verdin, M. E. Brown, M. Barlow, and A. Hoell. 2008. Warming of the Indian Ocean threatens eastern and southern African food security but could be mitigated by agricultural development. Proceedings of the National Academy of Sciences **105**:11081–11086.

Funk, C. C., and M. E. Brown. 2009. Declining global per capita agricultural production and warming oceans threaten food security. Food Security **1**:271–289.

Gallopin, G. 2006. Linkages between vulnerability, resilience, and adaptive capacity. Global Environmental Change **16**:293–303.

Game, E., M. Watts, S. Wooldridge, and H. Possingham. 2008a. Planning for persistence in marine reserves: a question of catastrophic importance. Ecological Applications **18**:670–680.

Game, E. T., M. Bode, E. McDonald-Madden, H. S. Grantham, and H. P. Possingham. 2009. Dynamic marine protected areas can improve the resilience of coral reef systems. Ecology Letters **12**:1336–1345.

Game, E. T., E. McDonald-Madden, M. L. Puotinen, and H. P. Possingham. 2008b. Should we protect the strong or the weak? Risk, resilience, and the selection of marine protected areas. Conservation Biology **22**:1619–1629.

Gardner, T. A., I. M. Cote, J. A. Gill, A. Grant, and A. R. Watkinson. 2003. Long-term region-wide declines in Caribbean corals. Science **301**:958–960.

Gardner, T. A., I. M. Cote, J. A. Gill, A. Grant, and A. R. Watkinson. 2005. Hurricanes and caribbean coral reefs: Impacts, recovery patterns, and role in long-term decline. Ecology **86**:174–184.

Gelcich, S., G. Edwards-Jones, M. J. Kaiser, and J. C. Castilla. 2006. Co-management policy can reduce resilience in traditionally managed marine ecosystems. Ecosystems **9**:951–966.

Gelcich, S., N. Godoy, L. Prado, and J. C. Castilla. 2008b. Add-on conservation benefits of marine territorial user rights fishery policies in Central Chile. Ecological Applications **18**:273–281.

Gelcich, S., M. Silva, M. Kaiser, and G. Edwards-Jones. 2008a. Engagement in co-management of marine benthic resources influences environmental perceptions of artisanal fishers. Environmental Conservation **35**:36–45.

Gell, F. R., and C. M. Roberts. 2003. Benefits beyond boundaries: the fishery effects of marine reserves. Trends in Ecology and Evolution **18**:448–455.

Gjertsen, H. 2005. Can habitat protection lead to improvements in human well-being? evidence from marine protected areas in the Philippines. World Development **33**:199–217.

Glantz, M. 1992. Climate Variability, Climate Change, and Fisheries. Cambridge University Press, Cambridge.

Glenn, R., and T. Pugh. 2006. Epizootic shell disease in American lobster (*Homarus americanus*) in Massachusetts coastal waters: interactions of temperature, maturity, and intermolt duration. Journal of Crustacean Biology **26**:639–645.

Glynn, P. W. 1991. Coral reef bleaching in the 1980s and possible connections with global warming. Trends in Ecology and Evolution **6**:175–179.

Glynn, P. W. 1993. Coral reef bleaching: ecological perspectives. Coral Reefs **12**:1–17.

Glynn, P. W. 1997. Bioerosion and coral-reef growth: a dynamic balance. Pages 68–113 *in* C. Birkeland, editor. Life and Death of Coral Reefs. Chapman & Hall, New York.

Glynn, P. W. 2000. El Nino-Southern Oscillation mass mortalities of reef corals: a model of high temperature marine extinctions? Pages 117–133 *in* E. Insalaco, P. W. Skelton, and T. J. Palmer, editors. Carbonate platform systems: components and interactions. Geological Society of London, London.

Glynn, P. W., S. B. Colley, J. H. Ting, J. L. Maté, and H. M. Guzmán. 2000. Reef coral reproduction in the eastern Pacific: Costa Rica, Panama, and Galápagos Islands (Ecuador) - IV. Agariciidae, recruitment and recovery of *Pavona varians* and *Pavona* sp. a. Marine Biology **136**:785–805.

Golbuu, Y., S. Victor, L. Penland, D. Idip Jr., C. Emaurois, K. Okaji, H. Yukihira, A. Iwase, and R. Van Woesik. 2007. Palau's coral reefs show differential habitat recovery following the 1998 bleaching event. Coral Reefs **26**:319–332.

Gong, D.-Y., and J. Luterbacher. 2008. Variability of the low-level cross-equatorial jet of the western Indian Ocean since 1660 as derived from coral proxies. Geophysical Research Letters **35**:L01705.

Gordon, H. S. 1954. The economic theory of a common-property resource: the fishery. Journal of Political Economics **62**:124–141.

Goreau, T., T. McClanahan, R. Hayes, and A. Strong. 2000. Conservation of coral reefs after the 1998 global bleaching event. Conservation Biology **14**:5–15.

Grafton, R. Q. 2010. Adaptation to climate change in marine capture fisheries. Marine Policy **34**:606–615.

Graham, E. M., A. H. Baird, and S. R. Connolly. 2008a. Survival dynamics of scleractinian coral larvae and implications for dispersal. Coral Reefs **27**:529–539.

Graham, N., S. Wilson, S. Jennings, N. Polunin, J. Robinson, and T. Daw. 2007. Lag effects in the impacts of the 1998 bleaching event on coral reef fisheries in the Seychelles. Conservation Biology **21**:1291–1300.

Graham, N. A. J. 2007. Ecological versatility and the decline of coral feeding fishes following climate driven coral mortality. Marine Biology **153**:119–127.

Graham, N. A. J., T. R. McClanahan, M. A. MacNeil, S. K. Wilson, N. V. C. Polunin, S. Jennings, P. Chabanet, S. Clark, M. D. Spalding, Y. Letourneur, L. Bigot, R. Galzin, M. C. Öhman, K. C. Garpe, A. J. Edwards, and C. R. C. Sheppard. 2008b. Climate warming, marine protected areas and the ocean-scale integrity of coral reef ecosystems. PLoS One **3**:e30309.

Graham, N. A. J., S. K. Wilson, S. Jennings, N. V. C. Polunin, J. Bijoux, and J. Robinson. 2006. Dynamic fragility of oceanic coral reef ecosystems. Proceedings of the National Academy of Sciences **103**:8425–8429.

Graham, N. A. J., S. K. Wilson, M. S. Pratchett, N. V. C. Polunin, and M. D. Spalding. 2009. Coral mortality versus structural collapse as drivers of corallivorous butterflyfish decline. Biodiversity and Conservation **18**:3325–3336.

Graham, T., N. Idechong, and K. Sherwood. 2000. The value of dive-tourism and the impact of coral bleaching on diving in Palau. Pages 59–71 *in* H. Schuttenberg, editor. Causes, Consequences and Response. Selected papers presented at the 9th international Coral Reef Symposium on "Coral bleaching: assessing and linking ecological and socioeconomic impacts, future trends and mitigation planning." Coastal Resources Center. University of Rhode Island, Narragansett.

Grandcourt, E. M., and H. Cesar. 2003. The bio-economic impact of mass coral mortality on the coastal reef fisheries of the Seychelles. Fisheries Research **60**:539–550.

Gregg, W. W., M. E. Conkright, P. Ginoux, J. E. O'Reilly, and N. W. Casey. 2003. Ocean primary production and climate: Global decadal changes. Geophysical Research Letters **30**:1809–1813.

Grothmann, T., and A. Patt. 2005. Adaptive capacity and human cognition: the process of individual adaptation to climate change. Global Environmental Change **15**: 199–213.

Gunderson, L. 1999. Resilience, flexibility and adaptive management—antidotes for spurious certitude? Conservation Ecology **3**:23–24.

Guzman, H. M., and J. Cortes. 2007. Reef recovery 20 years after the 1982–1983 El Nino massive mortality. Marine Biology **151**:401–411.

Hadley Centre. 2009. http://hadobs.metoffice.com/hadisst/HadISST.

Hagmann, J., and E. Chuma. 2002. Enhancing the adaptive capacity of the resource users in natural resource management. Agricultural Systems **73**:23–39.

Hahn, T., P. Olsson, C. Folke, and K. Johansson. 2006. Trust-building, knowledge generation and organizational innovations: The role of a bridging organization for adaptive comanagement of a wetland landscape around Kristianstad, Sweden. Human Ecology **34**:573–592.

Halford, A., A. J. Cheal, D. Ryan, and D. M. Williams. 2004. Resilience to large-scale disturbance in coral and fish assemblages on the great barrier reef. Ecology **85**:1892–1905.

Halpern, B. S., S. E. Lester, and J. B. Kellner. 2010. Spillover from marine reserves and the replenishment of fished stocks. Environmental Conservation **36**: 268–276.

Hara, M., and J. Raakjaer Nielsen. 2003. Experiences with fisheries co-management in Africa. Pages 81–97 *in* D. Wilson, J. Raakjaer Nielsen, and P. Degnbol, editors. The Fisheries Co-management Experience. Accomplishments, Challenges and Prospects. Kluwer, Dordrecht.

Hardin, G. 1968. The tragedy of the commons. Science **162**:1243–1248.

Hay, S. I., J. Cox, D. J. Rogers, S. E. Randolph, D. I. Stern, J. Cox, G. D. Shanks, and R. W. Snow. 2002. Climate change and malaria resurgence in East African highlands. Nature **415**:905–909.

Hilborn, R. 2007. Defining success in fisheries and conflicts in objectives. Marine Policy **31**:153–158.

Hilborn, R., and K. Stokes. 2010. Defining overfished stocks: have we lost the plot? Fisheries **35**:113–120.

Hixon, M. A., and B. A. Menge. 1991. Species diversity: Prey refuges modify the interactive effects of predation and competition. Theoretical Population Biology **39**:178–200.

Hodgson, G. 1999. A global assessment of human effects on coral reefs. Marine Pollution Bulletin **38**:345–355.

Holbrook, S. J., R. J. Schmitt, and A. J. Brooks. 2008. Resistance and resilience of a coral reef fish community to changes in coral cover. Marine Ecology Progress Series **371**:263–271.

Holling, C., F. Berkes, and C. Folke. 1998. Science, sustainability and resource management. Pages 342–362 *in* F. Berkes and C. Folke, editors. Linking Social and Ecological Systems: Management Practices and Social Mechanisms for Building Resilience. Cambridge University Press, Cambridge.

Holling, C. S. 1973. Resilience and stability in ecological systems. Annual Review of Ecology and Systematics **4**:1–23.

Hoorweg, J., N. Versleijen, B. Wangila, and D. A. 2008. Income diversification and fishing practices among artisanal fishers on the Malindi-Kilifi coast. Page 18 *in* Coastal Ecology Conference IV, Mombasa.

Hope, K. 2009. Climate change and poverty in Africa. International Journal of Sustainable Development and World Ecology **16**:451–461.

Horton, M., and J. Midddleton. 2000. The Swahil. Blackwell, Oxford, UK.

Hughes, T. P. 1994. Catastrophes, phase shifts, and large-scale degradation of a Caribbean coral reef. Science **265**:1547–1551.

Hughes, T. P., A. H. Baird, E. A. Dinsdale, N. A. Moltschaniwskyj, M. S. Pratchett, J. E. Tanner, and B. L. Willis. 2000. Supply-side ecology works both ways: the link between benthic adults, fecundity, and larval recruits. Ecology **81**:2241–2249.

Hughes, T. P., D. R. Bellwood, and S. R. Connolly. 2002. Biodiversity hotspots, centres of endemicity, and the conservation of coral reefs. Ecology Letters **5**:775–784.

Hughes, T. P., M. J. Rodrigues, D. R. Bellwood, D. Ceccarelli, O. Hoegh Guldberg, L. McCook, N. Moltschaniwskyl, M. S. Pratchett, R. S. Steneck, and B. Willis. 2007. Phase shifts, herbivory, and resilience of coral reefs to climate change. Current Biology **17**:360–365.

Human Development Report. 2007. http://hdr.undp.org/en/statistics/.

IFREMER. 1999. Evaluation of the Fisheries Agreements concluded by the European Community. Community contract no. 97/S 240–152919 OF 10 December 1997, IFREMER.

Ihara, C., Y. Kushnir, and M. A. Cane. 2008. Warming trends of the Indian Ocean SST and Indian Ocean dipole from 1880 to 2004. Journal of Climate **21**:2035–2046.

Ikiara, M., and J. Odink. 2000. Fishermen resistance to exit fisheries. Marine Resource Economics **14**:199–213.

IPCC. 2001. Climate change 2001: impacts, adaptation and vulnerability. Contribution of Working Group II to the Third Assessment Report of the Intergovernamental Panel on Climate Change, Cambridge University Press, Cambridge, UK.

ITC. 2007. (International Trade Center) Market New Service Spices. UNCTAD/WTO, Geneva, Switzerland.

Ives, A. R., and S. R. Carpenter. 2007. Stability and diversity of ecosystems. Science **317**: 58–62.

Jackson, J., M. Kirby, W. Berger, K. Bjorndal, L. Bostford, B. Bourque, R. Bradbury, R. Cooke, J. Erlandson, J. Estes, T. Hughes, S. Kidwell, C. Lange, H. Lenihan, J. Pandolfi, C. Peterson, R. Steneck, M. Tegner, and R. Warner. 2001. Historical overfishing and the recent collapse of coastal ecosystems. Science **293**:629–637.

Jackson, J. B. C. 2001. What was natural in the coastal oceans? Proceedings of the National Academy of Sciences **98**:5411–5418.

Jones, G. P., G. R. Almany, G. R. Russ, P. F. Sale, R. S. Steneck, M. J. H. van Oppen, and B. L. Willis. 2009. Larval retention and connectivity among populations of corals and reef fishes: history, advances and challenges. Coral Reefs **28**:307–325.

Jones, G. P., M. I. McCormick, M. Srinivasan, and J. V. Eagle. 2004. Coral decline threatens fish biodiversity in marine reserves. Proceedings of the National Academic Sciences **101**:8251–8258.

Jury, M., T. McClanahan, and J. Maina. 2010a. West Indian Ocean variability and East African fish catch. Marine Environmental Research **70**:162–170.

Jury, C. P., R. F. Whitehead, and A. M. Szmant. 2010b. Effects of variations in carbonate chemistry on the calcification rate of *Madracis auretenra* (= *Madradis mirabilis* sensu Wells, 1973): bicarbonate concentrations best predict calcifiction rates. Global Change Biology **16**:1632–1644.

Kaczynski, V. M., and D. L. Fluharty. 2002. European policies in West Africa: who benefits from fisheries agreements? Marine Policy **26**:75–93.

Kaufmann, D., A. Kraay, and M. Mastruzzi. 2005. Governance matters IV: governance indicators for 1996–2005. World Bank Policy Research Working Paper 3630.

Kaunda-Arara, B., and G. A. Rose. 2006. Growth and survival rates of exploited coral reef fishes in Kenyan marine parks derived from tagging and length-frequency data. Western Indian Ocean Journal of Marine Science **5**:17–26.

Kayanne, H., H. Ijima, N. Nakamura, T. R. McClanahan, S. Behera, and Y. Yamagata. 2006. Indian ocean dipole index recorded in Kenyan coral annual density bands. Geophysical Research Letters **33**:10–14.

Kleypas, J. A., R. A. Feely, V. J. Fabry, C. Langdon, C. L. Sabine, and L. L. Robbins. 2006. Impacts of Ocean Acidification on Coral Reefs and Other Marine Calcifiers: A Guide for Future Research. Sponsored by NSF, NOAA, USGS, St. Petersburg, Fl.

Knopf, B., K. Zickfeld, M. Flechsig, and V. Petoukhov. 2008. Sensitivity of the Indian monsoon to human activities. Advances in Atmospheric Sciences **25**:932–945.

Knowlton, N., and J. B. C. Jackson. 2008. Shifting baselines, local impacts, and global change on coral reefs. PLoS Biology **6**:e54.

Kobashi, T., J. P. Severinghaus, and J.-M. Barnola. 2008. 4 ± 1.5°C abrupt warming 11,270 yr ago identified from trapped air in Greenland ice. Earth and Planetary Science Letters **268**:397–407.

Kramer, R., S. Simanjuntak, and C. Liese. 2002. Migration and fishing in Indonesian coastal villages. Ambio **31**:367–372.

Kropelin, S., D. Verschuren, A. M. Lezine, H. Eggermont, C. Cocquyt, P. Fracus, J. P. Cazet, M. Fagot, B. Rumes, J. M. Russell, F. Darius, D. J. Conley, M. Schuster, H. von Suchodoletz, and D. R. Engstrom. 2008. Climate-driven ecosystem succession in the Sahara: The past 6000 years. Science **320**:765–768.

Krishna, A. 2006. Pathways out of and into poverty in 36 villages of Andhra Pradesh, India. World Development **34**:271–288.

Krishna, A., P. Kritjanson, M. Radeny, and W. Nindo. 2004. Escaping poverty and becoming poor in 20 Kenyan villages. Journal of Human Development **5**:211–226.

Kuleshov, Y., L. Qi, R. Fawcett, and D. Jones. 2008. On tropical cyclone activity in the Southern hemisphere: trends and the ENSO connection. Geophysical Research Letters **35**:L14S08.

LaJeunesse, T. C., W. K. W. Loh, R. van Woesik, O. Hoegh-Guldberg, G. W. Schmidt, and W. K. Fitt. 2003. Low symbiont diversity in southern Great Barrier Reef corals, relative to those of the Caribbean. Limnology and Oceanography **48**:2046–2054.

LaJeunesse, T. C., D. J. Thornhill, E. F. Cox, F. G. Stanton, W. K. Fitt, and G. W. Schmidt. 2004. High diversity and host specificity observed among symbiotic dinoflagellates in reef coral communities from Hawaii. Coral Reefs **23**:596–603.

Lal, M. 2001. Tropical cyclones in a warmer world. Current Science **80**:1103–1104.

Lamb, H. F., C. R. Bates, P. V. Coombes, M. H. Marshall, M. Umer, S. J. Davies, and E. Dejen. 2007. Late Pleistocene desiccation of Lake Tana, source of the Blue Nile. Quaternary Science Reviews **26**:287–299.

Lamb, H. F., I. Darbyshire, and D. Verschuren. 2003. Vegetation response to rainfall variation and human impact in central Kenya during the past 1100 years. The Holocene **13**:285–292.

Lambin, E. 2007. The Middle Path: Avoiding Environmental Catastrophe. University of Chicago Press, Chicago.

Langdon, C., and M. J. Atkinson. 2005. Effect of elevated pCO2 on photosynthesis and calcification of corals and interactions with seasonal change in temperature/irradiance and nutrient enrichment. Journal of Geophysical Research **110,** C09S07:doi: 10.1029/2004JC002576.

Lantz, V., and R. Martinez-Espineira. 2008. Testing the environmental Kuznets curve hypothesis with bird populations as habitat-specific environmental indicators: evidence from Canada. Conservation Biology **22**:428–438.

Lebel, L., M. Anderies, B. Campbell, C. Folke, S. Hatfield-Dodds, T. Hughes, and J. Wilson. 2006. Governance and the capacity to manage resilience in regional social-ecological systems. Ecology and Society **11**:19.

Lester, S. E., B. S. Halpern, K. Grorud-Colvert, J. Lubchenco, B. I. Ruttenberg, S. D. Gaines, S. Airame, and R. R. Warner. 2009. Biological effects within no-take marine reserves: a global synthesis. Marine Ecology Progress Series **384**:33–46.

Letourneur, Y., M. Harmelin-Vivien, and R. Galzin. 1993. Impact of hurricane Firinga on fish community structure on fringing reefs of Reunion Island, S.W. Indian Ocean. Environmental Biology of Fishes **37**:109–120.

Liu, Y., and U. R. Sumaila. 2008. Can farmed salmon production keep growing? Marine Policy **32**:497–501.

Livingston, M., and K. Sullivan. 2007. Successes and challenges in the hoki fishery of New Zealand. Chapter 12 *in* T. McClanahan and J. Castilla, editors. Fisheries Management: Progress towards Sustainability. Blackwell, London.

LMMA. 2009. Locally managed marine area network. http://www.lmmanetwork.org/.

Lokrantz, J., M. Nystrom, M. Thyresson, and C. Johansson. 2008. The non-linear relationship between body size and function in parrotfishes. Coral Reefs **27**:967–974.

Loya, Y., K. Sakai, K. Yamazato, Y. Nakano, H. Sambali, and R. van Woesik. 2001. Coral bleaching: the winners and the losers. Ecology Letters **4**:122–131.

Macdonald, R. W., T. Harner, and J. Fyfe. 2005. Recent climate change in the Arctic and its impact on contaminant pathways and interpretation of temporal trend data. Science of the Total Environment **342**:5–86.

Macfayden, G., and E. Allison. 2009. Climate Change, Fisheries, Trade and Competitiveness: Understanding Impacts and Formulating Responses for Commonwealth Small States. Commonwealth Secretariat, London.

Magadza, C. 2000. Climate change impacts and human settlements in Africa: prospects for adaptation. Environmental Monitoring and Assessment **61**:193–205.

Maimbo, S. 2006. Remittances and economic development in Somalia. Paper 38. World Bank, Washington, DC.

Maina, J., V. Venus, T. R. McClanahan, and M. Ateweberhan. 2008. Modelling susceptibility of coral reefs to environmental stress using remote sensing data and GIS models in the western Indian Ocean. Ecological Modelling **212**:180–199.

Makumbe, J. 1998. Is there a civil society in Africa? International Affairs **74**:305–317.

Mangi, S. C., and C. M. Roberts. 2006. Quantifying the environmental impacts of artisanal fishing gear on Kenya's coral reef ecosystems. Marine Pollution Bulletin **52**:1646–1660.

Mann, M. E., Z. Zhang, S. Rutherford, R. S. Bradley, M. K. Hughes, D. Shindell, C. Ammann, G. Faluvegi, and F. Ni. 2009. Global signatures and dynamical origins of the Little Ice Age and Medieval climate anomaly. Science **326**:1256–1260.

Manzello, D. P., M. Brandt, T. B. Smith, D. Lirman, J. C. Hendee, and R. S. Nemeth. 2007. Hurricanes benefit bleached corals. Proceedings of the National Academy of Sciences **105**:12035–12039.

Marcus, R. 2001. Seeing the forest for the trees: integrated conservation and development projects and local perceptions of conservation in Madagascar. Human Ecology **29**:381–397.

Marshall, N. A., and P. A. Marshall. 2007. Conceptualizing and operationalizing social resilience within commercial fisheries in northern Australia. Ecology and Society **12**:1.

Marshall, P. A., and A. H. Baird. 2000. Bleaching of corals on the Great Barrier Reef: differential susceptibilities among taxa. Coral Reefs **19**:155–163.

Mascia, M. B., and C. A. Claus. 2009. A property rights approach to understanding human displacement from protected areas: the case of marine protected areas. Conservation Biology **23**:16–23.

May, R. M., J. R. Beddington, C. W. Clark, S. J. Holt, and R. M. Laws. 1979. Management of multispecies fisheries. Science **205**:267–277.

McClanahan, T. R. 1995. A coral reef ecosystem-fisheries model: impacts of fishing intensity and catch selection on reef structure and processes. Ecological Modelling **80**:1–19.

McClanahan, T. R. 2000a. Recovery of the coral reef keystone predator, *Balistapus undulatus*, in East African marine parks. Biological Conservation **94**:191–198.

McClanahan, T. R. 2000b. Bleaching damage and recovery potential of Maldivian coral reefs. Marine Pollution Bulletin **40**:587–597.

McClanahan, T. 2008. Response of the coral reef benthos and herbivory to fishery closure management and the 1998 ENSO disturbance. Oecologia **155**:169–177.

McClanahan, T. R. 2002a. A comparison of the ecology of shallow subtidal gastropods between western Indian Ocean and Caribbean coral reefs. Coral Reefs **21**:399–406.

McClanahan, T. R. 2002b. The near future of coral reefs. Environmental Conservation **29**:460–483.

McClanahan, T. R. 2007. Management of area and gear in Kenyan coral reefs. Pages 166–185 *in* T. R. McClanahan and J. C. Castilla, editors. Fisheries Management: Progress towards Sustainability. Blackwell, London.

McClanahan, T. R. 2010. Effects of fisheries closures and gear restrictions on fishing income in a Kenyan coral reef. Conservation Biology **24**: 1519–1528.

McClanahan, T., M. Ateweberhan, C. Sebastian, N. Graham, S. Wilson, M. Guillaume, and J. Bruggemann. 2007a. Western Indian Ocean coral communities: bleaching responses and susceptibility to extinction. Marine Ecology Progress Series **337**:1–13.

McClanahan, T. R., M. Ateweberhan, C. A. Muhando, J. Maina, and M. S. Mohammed. 2007b. Effects of climate and seawater temperature variation on coral bleaching and mortality. Ecological Monographs **77**:503–525.

McClanahan, T., J. Castilla, A. White, and O. Defeo. 2008a. Healing small-scale fisheries and enhancing ecological benefits by facilitating complex social-ecological systems. Reviews in Fish Biology and Fisheries **19**:33–47.

McClanahan, T., and J. Cinner. 2008. A framework for adaptive gear and ecosystem-based management in the artisanal coral reef fishery of Papua New Guinea. Aquatic Conservation: Marine and Freshwater Ecosystems **18**:493–507.

McClanahan, T. R., J. E. Cinner, J. Maina, N. A. J. Graham, T. M. Daw, S. M. Stead, A. Wamukota, K. Brown, M. Ateweberhan, V. Venus, and N. V. C. Polunin. 2008b. Conservation action in a changing climate. Conservation Letters **1**:53–59.

McClanahan, T. R., J. Cinner, A. T. Kamukuru, C. Abunge, and J. Ndagala. 2009e. Management preferences, perceived benefits and conflicts among resource users and managers in the Mafia Island Marine Park, Tanzania. Environmental Conservation **35**:340–350.

McClanahan, T. R., J. E. Cinner, N. A. J. Graham, T. M. Daw, J. Maina, S. M. Stead, A. Wamukota, K. Brown, V. Venus, and N. V. C. Polunin. 2009d. Identifying reefs of hope and hopeful actions: contextualizing environmental, ecological, and social parameters to respond effectively to climate change. Conservation Biology **23**:662–671.

McClanahan, T., H. Glaesel, J. Rubens, and R. Kiambo. 1997. The effects of traditional fisheries management on fisheries yields and the coral-reef ecosystems of southern Kenya. Environmental Conservation **24**:105–120.

McClanahan, T., N. Graham, J. Calnan, and M. MacNeil. 2007b. Towards pristine biomass: reef fish recovery in coral reef marine protected areas in Kenya. Ecological Applications **17**:1055–1067.

McClanahan, T. R., N. A. J. Graham, S. K. Wilson, Y. Letourneur, and R. Fisher. 2009c. Effects of fisheries closure size, age, and history of compliance on coral reef fish communities in the western Indian Ocean. Marine Ecology Progress Series **396**:99–109.

McClanahan, T., C. Hicks, and S. Darling. 2008c. Malthusian overfishing and efforts to overcome it on Kenyan coral reefs. Ecological Applications **18**:1516–1529.

McClanahan, T., J. Maina, R. Moothien Pillay, and A. Baker. 2005a. Effects of geography, taxa, water flow, and temperature variation on coral bleaching intensity in Mauritius. Marine Ecology Progress Series **298**:131–142.

McClanahan, T., J. Maina, and L. Pet-Soede. 2002. Effects of the 1998 coral morality event on Kenyan coral reefs and fisheries. Ambio **31**:543–550.

McClanahan, T., N. Muthiga, and S. Mangi. 2001. Coral and algal changes after the 1998 coral bleaching: Interaction with reef management and herbivores on Kenyan reefs. Coral Reefs **19**:380–391.

McClanahan, T. R., and J. Maina. 2003. Response of coral assemblages to the interaction between natural temperature variation and rare warm-water events. Ecosystems **6**:551–563.

McClanahan, T. R., J. Maina, and J. Davies. 2005c. Perceptions of resource users and managers towards fisheries management options in Kenyan coral reefs. Fisheries Management and Ecology **12**:105–112.

McClanahan, T. R., J. Maina, C. J. Starger, P. Herron-Perez, and E. Dusek. 2005b. Detriments to post-bleaching recovery of corals. Coral Reefs **24**:230–246.

McClanahan, T. R., and S. Mangi. 2000. Spillover of exploitable fishes from a marine park and its effect on the adjacent fishery. Ecological Applications **10**:1792–1805.

McClanahan, T. R., and S. Mangi. 2004. Gear-based management of a tropical artisanal fishery based on species selectivity and capture size. Fisheries Management and Ecology **11**:51–60.

McClanahan, T. R., M. J. Marnane, J. E. Cinner, and W. E. Kiene. 2006. A comparison of marine protected areas and alternative approaches to coral-reef management. Current Biology **16**:1408–1413.

McClanahan, T. R., and J. C. Mutere. 1994. Coral and sea urchin assemblage structure and interrelationships in Kenyan reef lagoons. Hydrobiologia **286**:109–124.

McClanahan, T. R., N. A. Muthiga, J. Maina, A. T. Kamukuru, and S. A. S. Yahya. 2009b. Changes in northern Tanzania coral reefs during a period of increased fisheries management and climatic disturbance. Aquatic Conservation: Marine and Freshwater Ecosystems 19:758–771.

McClanahan, T. R., E. Weil, and J. Maina. 2009a. Strong relationship between coral bleaching and growth anomalies in massive *Porites*. Global Change Biology 15:1804–1816.

McClanahan T. R., N. A. J. Graham, M. A. MacNeil, N. A. Muthiga, J. E. Cinner, J. H. Bruggemann, S. K. Wilson. in press. Critical thresholds and tangible targets for ecosystem-based fisheries management of coral reefs. Proceedings of the National Academy of Sciences.

McClennen, C. 2006. Shrimp disease and sustainability: status and consequences for farmed shrimp. Pages 131–146 *in* A. Rahman, A. H. G. Quddus, B. Pokrant, and A. M. Liaquat, editors. Shrimp Farming and Industry: Sustainability Trade and Livelihoods. University Press Ltd, Bangladesh.

McCook, L. J., T. Ayling, M. Cappo, J. H. Choat, R. D. Evans, D. M. De Freitas, M. Heupel, T. P. Hughes, G. P. Jones, B. Mapstone, H. Marsh, M. Mills, F. J. Molloy, C. R. Pitcher, R. L. Pressey, G. R. Russ, S. Sutton, H. Sweatman, R. Tobin, D. R. Wachenfeld, and D. H. Williamson. 2010. Adaptive management of the Great Barrier Reef: A globally significant demonstration of the benefits of networks of marine reserves. Proceedings of the National Academic Sciences **107**:18278-18285.

McIlwain, J. L., and G. P. Jones. 1997. Prey selection by an obligate coral-feeding wrasse and its response to small-scale disturbance. Marine Ecology Progress Series **155**:189–198.

Mora, C. 2008. A clear human footprint in the coral reefs of the Caribbean. Proceedings of the Royal Society B Biological Sciences **275**:767–773.

Mortimore, M. 1989. Adapting to drought: farmers, famines and desertification in West Africa. Cambridge University Press, Cambridge.

Mumby, P., J. Chisholm, A. Edwards, C. Clark, E. Roark, S. Andrefouet, and J. Jaubert. 2001. Unprecedented bleaching-induced mortality in *Porites* spp. at Rangiroa Atoll, French Polynesia. Marine Biology **139**:183–189.

Mumby, P., C. Dahlgren, A. Harborne, C. Kappel, F. Micheli, D. Brumbaugh, K. Holmes, J. Mendes, K. Broad, J. Sanchirico, K. Buch, S. Box, R. Stoffle, and A. Gill. 2006. Fishing, trophic cascades, and the process of grazing on coral reefs. Science **311**:98–101.

Mumby, P., A. Harborne, J. Williams, C. Kappel, D. Brumbaugh, F. Micheli, K. Holmes, C. Dahlgren, C. Paris, and P. Blackwell. 2007. Trophic cascade facilitates coral recruitment in a marine reserve. Proceedings of the National Academy of Science **104**:8362–8367.

Munday, P. L., G. P. Jones, M. S. Pratchett, and A. J. Williams. 2008. Climate change and the future for coral reef fishes. Fish and Fisheries **9**:261–285.

Munday, P. L., N. E. Crawley, and G. E. Nilsson. 2009a. Interacting effects of elevated temperature and ocean acidification on the aerobic performance of coral reef fishes. Marine Ecology Progress Series **388**:235–242.

Munday, P. L., J. M. Donelson, D. L. Dixson, and G. G. K. Endo. 2009b. Effects of ocean acidification on the early life history of a tropical marine fish. Proceedings of the Royal Society Biological Sciences **276**:3275–3283.

Munday, P. L., D. L. Dixson, J. M. Donelson, G. P. Jones, M. S. Pratchett, G. V. Devitsina, and K. B. Doving. 2009c. Ocean acidification impairs olfactory discrimination and homing ability of a marine fish. Proceedings of the National Academy of Sciences **106**:1848–1852.

Myers, N., R. A. Mittermeier, C. G. Mittermeier, G. A. B. da Fonseca, and J. Kent. 2000. Biodiversity hotspots for conservation priorities. Nature **403**:853–858.

Nakamura, N., H. Kayanne, H. Iijima, T. R. McClanahan, S. K. Behera, and T. Yamagata. 2009. Mode shift in the Indian Ocean climate under global warming stress. Geophysical Research Letters **36**:L23708.

Nakamura, T., and R. van Woesik. 2001. Water-flow rates and passive diffusion partially explain differential survival of corals during the 1998 bleaching event. Marine Ecology Progress Series **212**:301–304.

Naylor, R. L., R. J. Goldbury, A. Price, J. H. Primavera, N. Kautsky, M. C. Beveridge, J. Clay, C. Folke, J. Lubchenco, H. Mooney, and M. Troell. 2000. Effect of aquaculture on world fish supplies. Nature **405**:1017–1024.

Nicholls, R., and J. Lowe. 2004. Benefits of mitigation of climate change for coastal areas. Global Environmental Change **14**:229–244.

Nilsson, G. E., N. Crawley, I. G. Lunde, and P. L. Munday. 2009. Elevated temperature reduces the respiratory scope of coral reef fishes. Global Change Biology **15**:1405–1412.

Njock, J. C., and L. Westlund. 2010. Migration, resource management and global change: Experiences from fishing communities in West and Central Africa Marine Policy **34**:752–760.

NOAA. 2009. Fisheries of the United State 2008. NOAA, Maryland.

Norris, P., and R. Inglehart. 2004. Sacred and secular: religion and politics worldwide. Cambridge University Press, Cambridge.

Norstrom, A. V., M. Nystrom, J. Lokrantz, and C. Folke. 2009. Alternative states on coral reefs: beyond coral-macroalgal phase shifts. Marine Ecology Progress Series **376**:295–306.

Odum, E. P. 1985. Trends expected in stressed ecosystems. Bioscience **35**:419–422.

OECD. 2007. http://www.oecd.org/dataoecd/12/25/40578365.pdf.

OECD. 2010. Organization for Economic Cooperation and Development Statistical Extracts. http://stats.oecd.org/Index.aspx.

Ogallo, L. J. 1989. The spatial and temporal patterns of the east African seasonal rainfall derived from principal component analysis. International Journal of Climatology **9**:145–167.

Oliver, J. K., R. Berkelmans, and C. M. Eakin. 2009. Coral bleaching in space and time. Pages 21–39 *in* M. van Oppen and J. M. Lough, editors. Coral Bleaching: Patterns, Processes, Causes and Consequences. Springer Ecological Studies, Berlin.

Olsson, P., C. Folke, and F. Berkes. 2004. Adaptive comanagement for building resilience in social-ecological systems. Environmental Management **34**:75–90.

Ostrom, E. 1990. Governing the commons: the evolution of institutions for collective action. Cambridge Univerisity Press, Cambridge.

Ostrom, E. 2000. Collective action and the evolution of social norms. Journal of Economic Perspectives **14**:137–158.

Ostrom, E., J. Burger, C. B. Field, R. B. Norgaard, and D. Policansky. 1999. Sustainability—Revisiting the commons: Local lessons, global challenges. Science **284**:278–282.

Overpeck, J., D. Rind, A. Lacis, and R. Healy. 1996. Possible role of dust-induced regional warming in abrupt climate change during the last glacial period. Nature **384**:447–449.

Paavola, J., and W. N. Adger. 2006. Fair adaptation to climate change. Ecological Economics **56**:594–609.

Parry, M. L., O. F. Canziani, J. P. Palutikof, and co-authors. 2007. Technical Summary. Pages 23–78 *in* M. L. Parry, O. F. Canziani, J. P. Palutiko, P. J. van der Linden, and C. E. Hanson, editors. Climate Change 2007: Impacts, Adaptation and Vulnerability. Contribution of Working Group II to the Fourth Assessment Report of the Intergovernmental Panel on Climate Change. Cambridge University Press, Cambridge, UK.

Pauly, D. 2008. Global fisheries: A brief review. Journal of Biological Research **9**:3–9.

Pauly, D., R. Watson, and J. Alder. 2005. Global trends in world fisheries: impacts on marine ecosystems and food security. Philosophical Transactions of the Royal Society B-Biological Sciences **360**:5–12.

Peters, R. 1998. The justification of education. Pages 207–230 *in* P. Hirst and P. White, editors. Philosophy and education. Routledge, London.

Pfeiffer, M., O. Timm, W.-C. Dullo, and D. Garbe-Schoenberg. 2006. Paired coral Sr/Ca and d18O records from the Chagos Archipelago: Late twentieth century warming affects rainfall variability in the tropical Indian Ocean. Geology **34**:1069–1072.

Pistorius, P. A., and F. E. Taylor. 2009. Declining catch rates of reef fish in Aldabra's marine protected areas. Aquatic Conservation: Marine and Freshwater Ecosystems **19**: S2–S9.

Pitcher, T., and M. Lam. 2010. Fishful thinking: rhetoric, reality, and the sea before us. Ecology & Society **15**:25.

Pitcher, T. J., D. Kalikoski, G. Pramod, and K. Short. 2009. Not honouring the code. Nature **457**:658–659.

Planes, S., G. P. Jones, and S. R. Thorrold. 2009. Larval dispersal connects fish populations in a network of marine protected areas. Proceedings of the National Academy of Sciences **106**:5693–5697.

Pollnac, R., B. Crawford, and L. Gorospe. 2001. Discovering factors that influence the success of community-based marine protected areas in the Visayas, Philippines. Ocean and Coastal Management **44**:683–710.

Porter, G. 1997. Euro-African fishing agreements: Subsidizing overfishing in African waters. Pages 7–33 *in* S. Burns, editor. Subsidies and Depletion of World Fisheries: Case Studies. World Wildlife Fund, Washington, DC.

Pratchett, M., P. Munday, S. Wilson, N. Graham, T. Daw, S. Stead, A. Wamukota, K. Brown, M. Ateweberhan, V. Venus, and N. Polunin. 2008. Effects of climate-induced coral bleaching on coral-reef fishers: ecological and economic consequences. Oceanography and Marine Biology: An Annual Review **46**:251–296.

Pratchett, M. S., S. K. Wilson, and A. H. Baird. 2006. Declines in the abundance of Chaetodon butterflyfishes following extensive coral depletion. Journal of Fish Biology **69**:1–12.

Ramanathan, V., C. Chung, D. Kim, T. Bettge, L. Buja, J. T. Kiehl, W. M. Washington, Q. Fu, D. R. Sikka, and M. Wild. 2005. Atmospheric brown clouds: impacts on South Asian climate and hydrological cycle. Proceedings of the National Academy of Sciences **102**:5326–5333.

Rapport, D. J., and W. G. Whitford. 1999. How ecosystems respond to stress. BioScience **49**:193–203.

Reason, C. J. C., and M. Rouault. 2002. ENSO-like decadal variability and South African rainfall. Geophysical Research Letters **29**:1638.

Reuters. 2007. Madagascar: successive cyclones bring Madagascar to its knees.

Richardson, A. J., and E. S. Poloczanska. 2008. Under-resourced, under threat. Science **320**:1294–1295.

Roberts, C. M., and J. P. Hawkins. 1999. Extinction risk in the sea. Ecology & Evolution **14**:241–246.

Roberts, C. M., and R. F. G. Ormond. 1987. Habitat complexity and coral reef fish diversity and abundance on Red Sea fringing reefs. Marine Ecology Progress Series **41**:1–8.

Rodrigues, L. J., and A. G. Grottoli. 2006. Calcification rate and the stable carbon, oxygen, and nitrogen isotopes in the skeleton, host tissue, and zooxanthellae of bleached and recovering Hawaiian corals. Geochimica Et Cosmochimica Acta **70**:2781–2789.

Rodriguez-Lanetty, M., S. Harii, and O. Hoegh-Guldberg. 2009. Early molecular responses of coral larvae to hyperthermal stress. Molecular Ecology **18**:5101–5114.

Rodwell, L., E. B. Barbier, C. M. Robert, and T. R. McClanahan. 2002. A model of tropical marine reserve-fishery linkages. Natural Resource Modeling **15**:453–486.

Rosenfeld, D. 1999. TRMM observed first direct evidence of smoke from forest fires inhibiting rainfall. Geophysical Research Letters **26**:3105–3108.

Rosenfeld, D., U. Lohmann, G. B. Raga, C. D. O'Dowd, M. Kulmala, S. Fuzzi, A. Reissell, and M. O. Andreae. 2008. Flood or drought: How do aerosols affect precipitation? Science **321**:1309–1313.

Ruud, J. 1960. Taboo: a study of Malagasy customs and beliefs. Oslo University Press, Oslo, Norway.

Sadovy, Y. J., and A. C. J. Vincent. 2002. Ecological issues and the trades in live reef fishes. Pages 391–340 *in* P. F. Sale, editor. Coral Reef Fishes: Dynamics and Diversity in a Complex Ecosystem. Academic Press, San Diego.

Saji, N., B. Goswami, P. Vinayachandran, and T. Yamagata. 1999. A dipole mode in the tropical Indian Ocean. Nature **401**:360–363.

Sanderson, E., M. Jaiteh, M. Levy, K. Redford, A. Wannebo, and G. Woolmer. 2002. The human footprint and the last of the wild. Bioscience **52**:891–904.

Sandin, S., J. Smith, E. DeMartini, E. Dinsdale, S. Donner, A. Friedlander, T. Konotchick, M. Malay, J. Maragos, D. Obura, O. Pantos, G. Paulay, M. Richie, F. Rohwer, R. Schroeder, S. Walsh, J. Jackson, N. Knowlton, and E. Sala. 2008. Baselines and degradation of coral reefs in the Northern Line Islands. PLoS One **3**:e1548.

Sano, Y. 2008. The role of social capital in a common property resource system in coastal areas: A case study of community-based coastal resource management in Fiji. SPC Traditional Marine Resource Management and Knowledge Information Bulletin **24**:19–32.

Satria, A., Y. Matsuda, and M. Sano. 2006. Questioning community based coral reef management systems: case study of Awig-Awig in Gili Indah, Indonesia. Environment, Development, and Sustainability **8**:99–118.

Scales, H., A. Balmford, M. Liu, Y. Sadovy, and A. Manica. 2006. Keeping Bandits at Bay? Science **313**:612–613.

Scales, H., A. Balmford, and A. Manica. 2007. Impacts of the live reef fish trade on populations of coral reef fish off northern Borneo. Proceedings of the Royal Society B **274**:989–994.

Schmitt, K., and D. Kramer. 2010. Road development and market access on Nicaragua's Atlantic coast: implications for household fishing and farming practices. Environmental Conservation **36**:289–300.

Schneider, S. 2009. Science as a Contact Sport. National Geographic, Washington, DC.

Selig, E. R., and J. F. Bruno. 2010. A global analysis of the effectiveness of marine protected areas in preventing coral loss. PLoS One **5**:e9278.

Selig, E. R., K. S. Casey, and J. F. Bruno. 2010. New insights into global patterns of ocean temperature anomalies: implications for coral reef health and management. Global Ecology and Biogeography **19**:397–411.

Semesi, I. S., S. Beer, and M. Bjork. 2009. Seagrass photosynthesis controls rates of calcification and photosynthesis of calcareous macroalgae in a tropical seagrass meadow. Marine Ecology Progress Series **382**:41–47.

Sheppard, C., A. Harris, and A. Sheppard. 2008. Archipelago-wide coral recovery patterns. Marine Ecology Progress Series **362**:109–117.

Sheppard, C., M. Spalding, C. Bradshaw, and S. Wilson. 2002. Erosion vs. recovery of coral reefs after 1998 EI Nino: Chagos reefs, Indian Ocean. Ambio **31**:40–48.

Sievanen, L., B. Crawford, R. Pollnac, and C. Lowe. 2005. Weeding through assumptions of livelihood approaches in ICM: Seaweed farming in the Philippines and Indonesia. Ocean & Coastal Management **48**:297–313.

Signa, and P. Tuda. 2008. BMU national workshop—opportunities and challenges for co-management. CORDIO, East Africa.

Silverman, J., B. Lazar, L. Cao, K. Caldeira, and J. Erez. 2009. Coral reefs may start dissolving when atmospheric CO_2 doubles. Geophysical Research Letters **36**:L05606.

Smit, B., and O. Pilifova. 2003. From adaptation to adaptive capacity and vulnerability reduction. Pages 9–28 *in* J. Smith, R. Klein, and S. Huq, editors. Climate Change, Adaptive Capacity and Development. Imperial College Press, London.

Smit, B., O. Pilifova, I. Burton, B. Challenger, S. Huq, R. Klein, and G. Yohe. 2001. Adaptation to Climate Change in the Context of Sustainable Development and Equity. Contribution of the Working Group II to the Third Assessment Report, Intergovernmental Panel on Climate Change. Cambridge University Press, Cambridge, UK.

Spalding, M., C. Ravilious, and E. P. Green. 2001. World Atlas of Coral Reefs. University of California Presa, Berkeley.

Stark, O. 1991. The Migration of Labour. Harvard University Press, Cambridge, MA.

Steffensen, J., K. Andersen, M. Bigler, H. Clausen, D. Dahl-Jensen, H. Fischer, K. Goto-Azuma, M. Hansson, S. Johnsen, J. Jouzel, V. Masson-Delmotte, T. Popp, S. Rasmussen, R. Rothlisberger, U. Ruth, B. Stauffer, M. Siggaard-Andersen, A. Sveinbjornsdottir, A. Svensson, and J. White. 2008. High-resolution Greenland ice core data show abrupt climate change happens in few years. Science **321**:680–684.

Steins, N. A., and V. M. Edwards. 1999. Collective action in common-pool resource management: The contribution of a social constructivist perspective to existing theory. Society & Natural Resources **12**:539–557.

Stenseth, N., A. Mysterud, G. Ottersen, J. Hurrell, K.-S. Chan, and M. Lima. 2002. Ecological effects of climate fluctuations. Science **297**:1292–1296.

Stillwell, J., A. Samba, P. Failler, and F. Laloe. 2010. Sustainable development consequences of European Union participation in Senegal's marine fishery. Marine Policy 34:616–623.

Stobart, B., K. Teleki, R. Buckley, N. Downing, and M. Callow. 2005. Coral recovery at Aldabra Atoll, Seychelles: Five years after the 1998 bleaching event. Philosophical Transactions of the Royal Society A: Mathematical, Physical and Engineering Sciences **363**:251–255.

Sumaila, R., J. Alder, and H. Keith. 2006. Global scope and economics of illegal fishing. Marine Policy **30**:696–703.

Sumaila, R., and M. Vasconcellos. 2000. Simulation of ecological and economic impacts of distant water fleets on Namibian fisheries. Ecological Economics **32**:457–464.

Suzuki, A., M. K. Gagan, K. Fabricius, P. J. Isdale, I. Yukino, and H. Kawahata. 2003. Skeletal isotope microprofiles of growth perturbations in *Porites* corals during the 1997–1998 mass bleaching event. Coral Reefs **22**:357–369.

Swartz, W., U. R. Sumaila, R. Watson, and D. Pauly. 2010. Sourcing seafood for the three major markets: The EU, Japan and the USA. Marine Policy **34**:1366–1373.

Tengo, M., and M. Hammer. 2003. Management practices for building adaptive capacity: a case from northern Tanzania. Pages 132–162 *in* F. Berkes, J. Colding, and C. Folke, editors. Navigating Social Ecological Systems: Building Resilience for Complexity and Change. Cambridge University Press, Cambridge, UK.

Thomas, C., B. Hewitson, A. Newsham, C. Twyman, and W. Adger. 2005. Adaptive: adaptations to climate change amongst natural resource-dependant societies in the developing world: across the Southern African climate gradient. Technical Reports 35, Tyndall Centre, Norwich, UK.

Thomas, D. T. C. 2005. Equity and justice in climate change adaptation amongst natural-resource-dependent societies. Global Environmental Change **15**:115–124.

Thompson, A. A., and A. M. Dolman. 2009. Coral bleaching: one disturbance too many for near-shore reefs of the Great Barrier Reef. Coral Reefs **29:** 637–648.

Thompson, D. W. J., J. J. Kennedy, J. M. Wallace, and D. Jones. 2008. A large discontinuity in the mid-twentieth century in observed global-mean surface temperature. Nature **453**:646–649.

Thompson, L. G., E. Mosley-Thompson, M. E. Davis, K. A. Henderson, H. H. Brecher, V. S. Zagorodnov, T. A. Mashiotta, P.-N. Lin, V. N. Mikhalenko, D. R. Hardy, and J. Beer. 2002. Kilimanjaro ice core records: evidence of Holocene climate change in tropical Africa. Science **298**:589–593.

Thompson, L. G., E. Mosley-Thompson, M. E. Davis, P. N. Lin, K. Henderson, and T. A. Mashiotta. 2003. Tropical glacier and ice core evidence of climate change on annual to millennial time scales. Climatic Change **59**:137–155.

Thompson, L. G., E. Mosley-Thompson, and K. Henderson. 2000. Ice-core palaeoclimate records in tropical South America since the last glacial maximum. Journal of Quaternary Science **15**:377–394.

Thornton, P., P. Jones, T. Owiyo, R. Kruska, M. Herrero, V. Orindi, S. Bhadwal, P. Kristjanson, A. Notenbaert, N. Bekele, and A. Omolo. 2008. Climate change and poverty in Africa: mapping hotspots of vulnerability. African Journal of Agricultural and Resource Economics **2**:24–44.

Tompkins, E., and W. Adger. 2004. Does adaptive management of natural resources enhance resilience to climate change? Ecology and Society **9**:10.

Tompkins, E., and W. Adger. 2005. Defining response capacity to enhance climate change policy. Environmental Science and Policy **8**:562–571.

Torell, E., J. Tobey, M. Thaxton, B. Crawford, B. Kalanghe, N. Madulu, A. Issa, V. Makota, and R. Sallema. 2006. Examining the linkages between AIDS and biodiversity conservation in coastal Tanzania. Ocean and Coastal Management **49**:792–811.

Trenberth, K. E. 2007. Warmer oceans, stronger hurricanes. Scientific American **297(1)**:45–51.

Tribollet, A., G. Decherf, P. A. Hutchings, and M. Peyrot-Clausade. 2002. Large-scale spatial variability in bioerosion of experimental coral substrates on the Great Barrier Reef (Australia): importance of microborers. Coral Reefs **21**:424–432.

Turner, J. R., E. Hardman, R. Klaus, I. Fagoonee, D. Daby, R. Baghooli, and S. Persands. 2000. The reefs of Mauritius. Pages 94–107 *in* D. Sounter, D. O. Obura , and O. Linden, editors. Coral Reef Degradation in the Indian Ocean: Status Report 2000. Stenby Offset, Vasteras, Sweden.

Turner, W. R., B. A. Bradley, L. D. Estes, D. G. Hole, M. Oppenheimer, and D. S. Wilcove. 2010. Climate change: helping nature survive the human response. Conservation Letters **3**:304-312.

Underwood, J. N., L. D. Smith, M. J. H. Van Oppen, and J. P. Gilmour. 2009. Ecologically relevant dispersal of corals on isolated reefs: Implications for managing resilience. Ecological Applications **19**:18–29.

UNEP. 2006. Annual Report. United Nations, Nairobi, Kenya. B2005.unstats.un.org/unsd/demographic/products/socind/default.htm.

Uyarra, M., I. Cote, J. Gill, R. Tinch, D. Viner, and A. Watkinson. 2005. Island-specific preferences of tourists for environmental features: implications of climate change for tourism-dependent states. Environmental Conservation **32**:11–19.

van Oppen, M. J. H., A. Baker, M. A. Coffroth, and B. L. Willis. 2009. Bleaching resistance and the role of algal endosymbionts. Pages 83–102 *in* M. J. H. van Oppen and J. M. Lough, editors. Coral Bleaching. Springer-Verlag, Berlin.

Verdin, J., C. Funk, G. Senay, and R. Choularton. 2005. Climate science and famine early warning. Philosophical Transactions of the Royal Society B: Biological Sciences **360**:2155–2168.

Vermeij, G. J. 1987. Evolution and Escalation: An Ecological History of Life. Princeton University Press, Princeton, NJ.

Verschuren, D., K. R. Laird, and B. F. Cumming. 2000. Rainfall and drought in equatorial east Africa during the past 1,100 years. Nature **403**:410–414.

Wallace, C. C., and H. Zahir. 2007. The "Xarifa" expedition and the atolls of the Maldives, 50 years on. Coral Reefs **26**:3–5.

Walley, C. 2004. Rough Waters: Nature and Development in an East African Marine Park. Princeton University Press, Princeton.

Walther, G., E. Post, P. Convey, A. Menzel, C. Parmesan, T. J. C. Beebee, J. Fromentin, O. Hoegh-Guldberg, and F. Bairlein. 2002. Ecological responses to recent climate change. Nature **416**:389–395.

Wang, Y., H. Cheng, R. L. Edwards, Y. He, X. Kong, Z. An, J. Wu, M. J. Kelly, C. A. Dykoski, and X. Li. 2005. The Holocene Asian monsoon: links to solar changes and North Atlantic climate. Science **308**:854–857.

Watson, R., and D. Pauly. 2001. Systematic distortions in world fisheries catch trends. Nature **414**:534–536.

Webster, P., G. Holland, J. Curry, and H.-R. Chang. 2005. Changes in tropical cyclone number, duration, and intensity in a warming environment. Science **309**:1844–1846.

Wegley, L., R. Edwards, B. Rodriguez-Brito, H. Liu, and F. Rohwer. 2007. Metagenomic analysis of the microbial community associated with the coral *Porites asteroides*. Environmental Microbiology **9**:2707–2719.

Wells, S. 2006. Assessing the effectiveness of marine protected areas as a tool for improving coral reef management. Pages 314–331 *in* I. M. Côté and J. D. Reynolds, editors. Coral Reef Conservation. Cambridge University Press, Cambridge, UK.

West, J., and R. Salm. 2003. Resistance and resilience to coral bleaching: implications for coral reef conservation and management. Conservation Biology **17**:956–967.

Western, D., S. Russell, and I. Cuthill. 2009. The status of wildlife in protected areas compared to non-protected areas in Kenya. PLoS One **4**:e6140.

Westmacott, S., H. Cesar, and L. Pet-Soede. 2000a. Socioeconomic assessment of the impacts of the 1998 coral reef bleaching in the Indian Ocean: a summary. Status Report 2000, CORDIO. SAREC Marine Science Program, Stockholm, Sweden.

Westmacott, S., H. Cesar, L. Pet Soede, and O. Linden. 2000b. Coral bleaching in the Indian Ocean: socio-economic assessment of effects. Pages 94–106 *in* H. Cesar, editor. Collected Essays on Economics of Coral Reefs. CORDIO, Kalmar, Sweden.

Westmacott, S., C. Teleki, S. Wells, and J. West. 2000c. Management of bleached and severely damaged coral reefs. IUCN, Gland, Switzerland.

Whitelaw, N., and E. Whitelaw. 2006. How lifetime shapes epigenotype with and across generations. Human Molecular Genetics **15**:R131–137.

Wickham, F., J. Kinch, and P. Lal. 2009. Institutional capacity within Melanesian countries to effectively respond to climate change impacts, with a focus on Vanuatu and the Solomon Islands. Secretariat of the Pacific Regional Environment Programme, Apia, Samoa.

Wilen, J. 2004. Spatial management of fisheries. Marine Resource Economics **19**:7–19.

Wilkinson, C., editor. 2000. Status of Corals of the World: 2000. Australian Institute of Marine Science, Townsville.

Williams, I. D., V. C. Polunin, and V. J. Hendrick. 2001. Limits to grazing by herbivorous fishes and the impact of low coral cover on macroalgal abundance on a coral reef in Belize. Marine Ecology Progress Series 222:187–196.

Wilson, S. K., A. M. Dolman, A. J. Cheal, M. J. Emslie, M. S. Pratchett, and H. P. A. Sweatman. 2009. Maintenance of fish diversity on disturbed coral reefs. Coral Reefs **28**:3–14.

Wilson, S. K., R. Fisher, M. S. Pratchett, N. A. J. Graham, N. K. Dulvy, R. A. Turner, A. Cakacakas, N. V. C. Polunin, and S. P. Rushton. 2008. Exploitation and habitat degradation as agents of change within coral reef fish communities. Global Change Biology **14**:2796–2809.

Wilson, S. K., N. A. J. Graham, M. S. Pratchett, G. P. Jones, and N. V. C. Polunin. 2006. Multiple disturbances and the global degradation of coral reefs: are reef fishes at risk or resilient? Global Change Biology **12**:2220–2234.

Wood, L., L. Fish, J. Laughren, and D. Pauly. 2008. Assessing progress towards global marine protection targets: shortfalls in information and action. Fauna & Flora International, Oryx, **42**:340–351.

World Bank. 2006. Taking Stock of Research on International Migration in Sub-Saharan Africa. World Bank, Washington, DC.

World Bank. 2010. <http://info.worldbank.org/governance/wgi/index.asp>.

World Resources Institute. 2003. http://earthtrends.wri.org/country_profiles/index.php.

Worm, B., E. B. Barbier, N. Beaumont, J. E. Duffy, C. Folke, B. S. Halpern, J. B. C. Jackson, H. K. Lotze, F. Micheli, S. R. Palumbi, E. Sala, K. A. Selkoe, J. J. Stachowicz, and R. Watson. 2006. Impacts of biodiversity loss on ocean ecosystem services. Science **314**:789–790.

Worm, B., R. Hilborn, J. K. Baum, T. A. Branch, J. S. Collie, C. Costello, M. J. Fogarty, E. A. Fulton, J. A. Hutchings, S. Jennings, O. P. Jensen, H. K. Lotze, P. M. Mace, T. R. McClanahan, C. Minto, S. R. Palumbi, A. M. Parma, D. Ricard, A. A. Rosenberg, R. Watson, and D. Zeller. 2009. Rebuilding global fisheries. Science **325**:578–585.

WTO. 2001. Africa. Vol. 1, WTO Press and Communications, Madrid, Spain.

Xu, L., S. Rondenay, and R. D. van der Hilst. 2007. Structure of the crust beneath the southeastern Tibetan Plateau from teleseismic receiver functions. Physics of the Earth and Planetary Interiors **165**:176–193.

York, R., E. Rosa, and T. Dietz. 2003. Footprints on the earth: the environmental consequence of modernity. American Sociological Review **68**:279–300.

Young, O. R. 2010. Institutional Dynamics: Emergent Patterns in International Environmental Governance. MIT Press, Cambride, MA.

Zhang, L., W. Dawes, and G. Walker. 2001. Response of mean annual evapotransportation to vegetation changes at catchment scale. Water Resources Research **37**:701–708.

Zinke, J., W. C. Dullo, G. A. Heiss, and A. Eisenhauer. 2004. ENSO and Indian Ocean subtropical dipole variability is recorded in a coral record off southwest Madagascar for the period 1659 to 1995. Earth and Planetary Science Letters **228**:177–194.

Zinke, J., M. Pfeiffer, O. Timm, W.-C. Dullo, and G. Brummer. 2009. Western Indian Ocean marine and terrestrial records of climate variability: a review and new concepts on land-ocean interactions since AD 1660. International Journal of Earth Sciences **98**:115–133.

Zinke, J., M. Pfeiffer, O. Timm, W.-C. Dullo, and G. J. A. Brummer. 2008. Western Indian Ocean marine and terrestrial records of climate variability: a review of the new concepts on land-ocean interactions since AD 1660. International Journal of Earth Science **98**:115–133.

INDEX

acclimation, 4, 37, 38, 44, 45, 52
accountability, 12, 15, 19, 31, 32, 67, 75, 76, 103, 112, 118, 123, 124, 126, 128
acidification, iv, 37, 51, 52, 57, 60, 69
Acropora, 46, 47,48, 54
adaptation, i, iii, iv, v, vii, viii, 1, 2, 3, 4, 5, 6,7,8, 10, 12, 13, 14, 16, 18, 20,21, 24, 26, 28,30,31, 32,34,36, 47, 38, 39, 40, 42,44, 45, 46, 48, 49,50, 52, 54, 56, 57, 58, 60, 62, 64, 66, 67, 68, 69, 70, 71, 72, 74, 75, 76, 77, 78, 79, 80, 81, 82, 83, 84, 85, 86, 87, 88, 89, 90, 91, 92, 93, 94, 95, 96, 97, 98, 99, 100, 101, 102, 103, 104, 105, 106, 108, 109, 110, 111, 112, 113, 114, 115, 116, 117, 118, 119, 123, 124, 125, 126, 127, 128, 120,121, 122, 123, 124, 125, 126,127, 128, 129, 130, 131, 132, 133, 134, 138, 140, 141, 142, 144, 146, 148, 149, 150, 151, 152
adaptive capacity, v, 5, 6, 13, 67, 68, 72, 76, 81, 82, 83, 88, 89, 90, 92, 96, 97, 99, 100,101, 102, 103, 104, 105, 106, 108, 109.110, 111, 112, 113, 114, 115, 116, 117, 118, 119, 120, 121, 122, 123, 125, 126, 127,128, 129, 130, 131, 132, 133, 138, 141, 149, 150, 151, 152, 154, 157, 162, 163, 165, 176, 177
adaptive management framework, 57
aerosol, 31
Africa, 29, 30, 31, 32, 34, 43, 44, 45, 50, 67, 69, 70, 75, 76, 78, 79, 81, 82, 85, 107, 117, 148, 150
agriculture, vii, xi, 3, 8, 14, 33, 69, 70, 85, 92, 115, 118, 119, 130 152
AHP, Xi, 103
AIDS, 3, 79, 177
Alaska, 21, 91
albedo, 24
Aldabra, 44, 50, 62, 176
algae, 39, 49, 50, 56, 60, 61, 62, 102, 129, 134, 135, 137, 138, 140, 143, 144, 146
alleles, 38
amplify, 3, 97, 98, 130, 151
Analytic Hierarchy Process, xi, 103
ancestor, 75
antagonist, 104, 125, 161
Antigua, 93
apartheid, 117, 122, 158
appliance, 83
aquaculture, 8, 9,10, 13, 18, 20, 67
aquarium trade, 65, 91
Arabia, 24, 31, 43, 44, 50
aragonite, 51, 52
arctic, 22

Asia, 7,15,17, 24, 25, 89, 128
asset, 72, 81, 82, 83,88, 89, 90 96, 97, 99, 102, 103, 104, 116, 117, 118, 119, 122, 130, 141, 150, 151
Atlantic, 24
atoll, 44, 45, 62, 86
Australia, 13, 14, 32, 41, 50, 144
axis, 6, 22, 43, 57, 100, 102, 104, 105, 113, 130, 136, 137, 138

Bamburi, 71, 74, 82, 104, 106
Barbuda, 93
barrier, 45, 54, 55, 61, 86, 94, 120, 129, 144, 147
batfish, 61
beach management units, xi, 75, 120, 123, 126, 148
beach seine, 72, 118, 131, 139, 145, 146
Belize, 93, 179
bilateral, 15
biodiversity, 55
bioerosion, 51
biogeography, 47
biologists, 4, 37
biomass, 14, 63, 66, 104, 105, 106, 112, 119, 130, 131, 132, 135, 136, 137, 138, 139, 141, 143, 148, 149, 151
black carbon, 31
bleach, 3, 37, 39, 40, 41, 42, 44, 45, 46, 47, 48, 49, 51, 52, 53, 54, 55, 56, 58, 59, 60, 61, 62, 64, 65, 67, 68, 69, 71, 80, 84, 87, 90, 100, 102, 106, 107, 108, 109, 110, 111, 112, 129, 140, 145, 146, 152
blueprint, 124, 151
BMU, xi, 75,120, 121, 123, 125, 126, 148
boat, 10, 93, 96, 119, 122, 131, 122
buoys, 25
butterflyfish, 59, 65, 105

calcification, 37, 39, 51, 52, 53, 135
calcium carbonate, 27, 37, 50
California, 13, 91, 176
Canada, 10, 19, 168
canning, 16
capital, 14, 15, 16, 19, 21, 76, 77, 108, 102, 103, 104, 114, 119, 120, 121, 122, 126, 145, 152
carbohydrate, 39
carbon, 4, 21, 31, 32, 34, 37, 52, 93
carbon dioxide, 21, 37, 52
carbon emissions, 21, 93
Caribbean, 41, 44, 45, 51, 53, 61,63, 92
cash, 73, 74, 80, 86, 97
caviar, 9
Chagos, 30, 44, 50, 54, 56
chemistry, 28, 51, 52,66

Chile, iv, 10, 19, 21, 104, 120, 147
China, 9, 10, 19, 21, 34, 25
cholera, 93
civil society, 76
climate change, iii, iv, v, vii, viii, xi, 1, 2,3, 4, 5, 6, 7, 10, 13, 14, 20, 21, 22, 23, 25, 27, 29, 30, 31, 32, 34, 35, 37, 39, 40, 41, 43, 45, 48, 49, 51, 52, 55, 56, 57, 59, 60, 61, 63, 64, 65, 66, 67, 68, 69, 71, 72, 74, 76, 83, 84, 85, 87, 88, 89, 91, 92, 93, 94, 98, 99, 100, 101, 102, 106, 110, 111, 112, 112, 113, 115, 120, 121, 122, 127, 128, 129, 130, 133, 140, 141, 145, 149, 150, 151, 152, 153
closure, 14, 56, 63, 94, 103, 106, 108, 109, 110, 111, 113, 120, 121, 128, 137, 138, 139, 140, 141, 142, 143, 145, 148, 149, 151
co-management, 121, 123, 124, 125, 126, 131
coastal erosion, 84, 85, 86, 93
Code of Conduct, 11, 12, 13, 14, 17, 18
coffee, 7
collapse, 1, 59, 62, 71, 84, 85, 137, 138, 146
collective , 75, 84, 122, 123, 124
commitment, 13, 122, 133
common-pool, 11, 79, 126, 176
community change, 38
community-based management, 75, 122, 125
Comoros, 17, 30, 44, 45, 50, 69, 75, 76, 78, 79, 81, 82
compliance, 10, 18, 111, 139, 142, 144, 147, 148, 152
construction, 4, 23, 27, 28, 86, 93
consumer, 10, 135
continental shelve, 20, 47
contract, 16, 17 18
cooperation, 78, 122, 126
cope, vi, 1, 3, 4, 68, 73, 81, 83, 85, 90, 97, 103, 115, 128
Copenhagen, 108
coral reef, v, 5, 35, 51, 57, 58, 59, 60, 61, 62, 63, 64, 65, 67, 90, 102, 103, 115, 118, 129, 131, 134, 135, 137, 138, 140, 142, 144, 145,146, 147, 151, 152, 153
corruption, viii, 5, 18, 75, 76, 122, 123, 126
crab, 93
credit, 89, 91, 97, 119
crises, 73, 157
crop, 32, 46, 47, 48, 53, 54, 67, 73, 74, 79, 85, 117, 128, 129
cyclone, 3, 5, 31, 34, 52, 53, 65, 67, 68, 69, 74, 84, 85, 86, 90, 93

dampen, 98, 130
Dar es Salaam, iv, 71, 74, 82, 104, 106
death, 40, 62, 117, 127
decision, viii, 55, 67, 76, 84, 94, 101, 152
degradation, 67, 69, 80, 86, 100, 151
degree heating weeks, 41, 42
delta, 86
demersal, 18

developed, 8, 9, 10, 12, 13, 14, 15, 16, 17, 18, 19, 20, 55, 67, 68, 80, 100, 102, 107 113, 120, 121, 125, 127, 134, 147, 151, 153
developing country, 15, 17,20
diet, 61, 63
dinoflagellate, 39
Diploastrea, 52
disaster, vi, 69, 76, 79, 89, 90, 93, 97, 122, 129, 132
disease, 1, 10, 45, 51, 58, 67, 79, 90, 93
distant-water, 13, 15, 16, 17, 21
disturbance, v, vii, 4, 5, 37, 38, 40, 41, 43, 45, 46, 47, 48, 49, 53, 54, 55, 56, 57, 59, 61, 62, 63. 65, 66, 67, 68, 69, 73, 74, 75, 76, 79, 80, 84, 85, 87, 88, 89, 91, 93, 94, 95, 97, 99, 100, 104, 105, 106, 108, 113, 128, 129, 134, 140, 141, 145, 146, 149, 152
dive, 87, 88
Doha Round, 15
donor, 17, 18, 74, 125, 126, 127, 133
downwelling, 26, 32
dune, 86

early warning , 69, 81, 84, 90, 91, 93, 111, 127, 137
eco-tourism, 110
economic, 84, 85, 87, 92, 95, 97, 111, 114, 116, 127, 129, 131, 132, 133, 141, 149, 150, 151
ecosystem, v, vii, viii, 1, 2, 3, , 4, 5, 7, 10, 11, 12, 13, 18, 21, 38, 40, 50, 54, 55, 56, 62, 66, 67, 80, 85, 88, 94, 98, 99, 100, 102, 104, 106, 108, 109, 110, 111, 112, 113, 116, 129, 130, 131, 134, 135, 136, 137, 141, 142, 144, 145, 146, 147, 149, 152, 153
ecosystem engineer, 110, 111
Ecuador, 10
education, 6, 80, 81, 90, 114, 116, 117, 122, 125, 127, 128
EEZ, XI, 11, 12, 15, 18, 21
efficiency, 20, 89, 91, 118, 132, 133
effluent, 10
effort, 11, 14, 17, 89, 90, 95, 98, 131, 147
El Niño Southern Oscillation, xi, 27, 28, 30, 45, 91, 135
emergency relief, 133
emission, 20, 32, 34, 52
emperor, 63, 143
employ, 16, 98
enforce, 11, 108, 123, 124
epigenetic, 38
equator, 20
Eritrea, , 1
Ethiopia, 1, 30, 32
ethnic, , 1, 132
Europe, 7, 14, 15, 24
European Common Fisheries Policy, 15
evolution, 4, 12, 37, 38, 51
excludability, 11
Exclusive economic zone, xi, 11, 20

export, , 7, 9, 10, 19, 91, 119
exposure, 5, 6, 32, 35, 37, 40, 49, 54, 55, 67, 68, 69, 94, 100, 102, 106, 107, 108, 109, 110, 111, 112, 113, 121, 128, 129, 133, 141, 151, 152
extinction, 47, 48, 140
extreme events, 20, 32, 34, 92

family planning, 118
family size, 117
FAO, 7, 8, 9, 10, 11, 12, 13, 14, 17, 18, 19, 91, 115
Favia, 47, 48
feedback, 31, 80, 121, 127, 130
fish, 7, 20, 65, 71, 91. 95, 104, 113, 131, 136, 149, 149
fisheries, 90, 125, 126, 137, 138, 139, 141, 149
fisheries yield, 14, 88, 97, 137, 139, 144
fishing, 77, 98, 110, 132, 135, 144, 147
flexibility, 72, 73, 89, 90, 119, 121
flood,
Food and Agriculture Organization, xi, 8, 14, 115
food security, 33, 84, 112, 141, 149
food web, 38, 39, 144
foreign fleet, 15, 16
forest, 25
framework, 121, 125, 133, 149, 151, 152, 153
France, iv, 17, 19
fuel, 13, 93, 94, 110
funeral, 117

gastropod, 147
gear, 71, 72, 73, 77, 88, 89, 91, 92, 93, 94, 95, 97, 98, 102, 104, 110, 118, 119, 120, 129, 130, 131, 135, 136, 138, 139, 140, 142, 144, 145, 146, 147, 148, 151
GELOSE Gestion Locale Sécurisée, xi, 103
gene, 4, 37, 38
generalist, 59, 65
geography, 20, 115
Germany, 13, 19
glacial, 22
global, iv, v, 7, 8, 9, 10, 11, 12, 13, 15, 17, 18, 19, 20, 21, 24, 25, 26, 27, 30, 32, 34, 40, 41, 45, 51, 52, 53, 59, 61, 67, 69, 94, 102, 108, 110, 122, 125, 139, 140, 151
global warming, 25, 52
goatfish, 63
governance, 122, 124, 126, 128, 152
Great Barrier Reef, 45, 54, 55, 61, 94, 144
Greater Horn, 30
greenhouse gas, vii, 4, 108
Greenland, 21, 22
gross domestic product, xi, 67, 70, 86
growth rate, 10, 78, 79

Gulf of Oman, 43, 44, 50
Gyrosmillia, 48

habitat, 9, 10, 47, 50, 57, 58, 61, 62, 63, 64, 65, 71, 75, 89, 105, 113, 128, 141, 145
Hadley Center, 25, 26, 29
harvest, 11, 131, 134, 138, 143
health, 62, 79, 82, 88, 90, 93, 117, 118, 128
heat stress, 100
herbivore, 58, 60, 61, 62, 63, 64, 65, 129, 135, 136, 137, 140,111, 122, 143,144
herbivory, 141, 144
hierarchy, xi, 24, 84, 103, 125, 133
history, 1, 5, 13, 17, 22, 38, 46, 47, 60, 84, 98, 127, 140, 151
HIV, 3, 79
household, 7, 67, 70, 71, 73, 74, 75, 80, 82, 83, 88, 89,91, 96, 97, 102, 115, 117, 118, 122, 127
human development index, xi, 82, 103
humanitarian, 74
hurricane, 53

ice, 22, 23, 24, 34, 93
illegal, 9, 15, 16, 18, 118, 126, 145
illness, 117
import, 18, 19
incentive, 17, 143
income, 74, 77, 83, 87, 89, 91, 96, 97, 109, 110, 118, 133, 141
India, 44, 45, 50, 77, 117
Indian Ocean, vii, ix, xi, 1, 2, 7, 16, 17, 18, 21, 22, 23, 24, 25, 26, 27, 28, 30, 31, 32, 33, 35, 40, 41, 42, 43, 44, 45, 46, 48, 49, 50, 52, 53, 56, 57, 63, 67, 69, 71, 74, 76, 78, 79, 80, 81, 82, 84, 85, 86, 87, 88, 100, 107, 109, 131, 134, 136, 137, 149
Indian Ocean dipole, xi, 27, 28, 30, 35, 40, 45
Indonesia, 19, 21, 75, 119, 141
institution, 116
insurance, 89, 91, 119
inter-tropical convergence zone, xi, 23
Intergovernmental Panel on Climate Change, xi, 2, 3, 20, 76, 86, 115
irrigation, 24, 84, 118
island nation, 31, 81
isotopes, 23, 24, 27, 29
Italy, iv, 17, 19

jacks, 65
Japan, iv, 15, 19
job, 16, 73, 74, 96, 102, 118
Johannesburg Plan, 12, 14

Kenya, 21, 24, 27, 30, 44, 45, 46
Kenya Wildlife Service, 121, 141
Kilimanjaro, 23, 24

Kuznet, 131
Kyoto, 108

Lake Malawi, 23
Lake Naivasha, 24
Lake Tanganyika, 91
Lake Victoria, 91, 97
larvae, 48, 54, 57, 59, 60, 139, 140
Latin America, 7
latitude, 20, 21, 24, 31, 48
learning, 72, 80, 89, 116, 120, 127, 128, 149, 150, 152
legal, 11, 12, 15, 76, 84, 152, 134
levees, 69
life span, 38
Little Ice Age, 24
loan, 117, 118
lobster, 9, 90
Locally Managed Marine Areas network, xi, 128
Lombok, 141

macroalgae, 58, 135, 136, 143
Madagascar, 17, 21, 28, 29, 30, 34, 44, 45, 49, 50,53, 69, 70, 71, 73, 74,75, 76, 77, 78, 79, 80, 81, 82, 83, 86, 92, 95, 96, 97, 101, 102, 103, 104, 106, 107, 108, 108, 109, 111, 113, 116, 120, 121, 123, 124, 131, 132, 142, 148, 149
Mafia, 27, 28, 30, 50, 142
Magnusson-Stevenson Act, 12
maladaptation, 97
malaria, 3, 84, 90, 93
Maldives, 44, 45, 50, 54, 69, 78, 79, 81, 82, 87, 137, 148, 151
management, xi, 6, 7, 11, 12, 13, 14, 15, 16, 17, 18, 19, 21, 36, 37, 43, 54
manager, 2, 5, 94, 114, 129, 134, 137, 138, 139, 140, 144, 149, 152
mangrove, 10, 63, 86
marine protected area, xi, 121
Marine Stewardship Council, 13
material style of life, 82, 82, 102, 119
Mauritius, 17, 44, 45, 50, 53, 69, 70-83, 95, 102-111, 132, 141, 147, 149
Mayotte, 29, 30, 43, 44, 45, 50, 148
Medieval Warm Period, 24
membership, 123, 124
mentor, 118
meta-population, 47
metabolism, 37
metric tons, 8, 14, 17, 19
migration, 23, 24, 57, 75, 78, 79, 80, 88, 91, 92, 121
minimum size, 144
mitigation, 4, 108, 115, 119
mobil, 77, 81, 90, 92, 93, 96, 99, 102, 104, 118, 120, 121

mollusk, 51, 93
Mombasa, 74, 87, 139, 142, 143
monitor, 122
monsoon, 24, 25, 27, 28
Montipora, 46, 47, 48
mortality, 14, 37, 41, 43, 44, 45, 46, 49, 51, 54, 55, 57, 59, 62, 63, 64, 65, 71, 84, 103, 105, 109, 128, 129, 135, 140, 145, 146, 146
multidisciplinary, 153
multispecies maximum sustained yield, xi, 137, 138, 139, 147, 148, 149
mutation, 38

Namibia, 17
National Adaptation Plans for Action, 115
National Oceanographic and Atmospheric Administration, xi, 25
neolithic, 24
New England, 13
Nicaragua, 119
NOAA, xi, 9, 25, 26
nongovernmental organization, xi, 12, 126
normalize, 29, 102, 103, 105
North America, 7
Norway, 10, 19, 21, 154
Nosy Antafana, 77, 116
nutrient, 39
nutrition, 52, 141

occupation, 70, 71, 74, 98, 102
oceanography, v, 23, 23, 25, 29, 31, 35
opportunity, 92, 151
Organization for Economic Co-operation and Development, xi, 18, 19, 78, 81
ornamental, 15, 65
oscillation, xi, 5, 27, 28, 30, 91
overexploitation, 1, 11, 86, 119, 120, 130, 147
overfishing, 9, 13, 14, 15, 51, 85, 108, 130
ownership, 17, 141, 147
oxygen, 23, 37
Oxypora, 48

Pacific, xi, 7, 15, 16, 30, 53, 61, 91, 128, 147
Pacific decadal oscillation, xi, 30
Papua New Guinea, 65, 75, 119, 143
parrotfish, 51, 61, 77, 105, 128, 144, 145
payoff, 101, 109
pelagic, 18, 28, 21, 31, 59, 65, 90, 110, 148
Persian Gulf, 43, 44, 50
perturbation, 4
Peru, 19

perverse, 11, 13, 74, 115, 132
pH, 51, 60
phase shift, 5, 134, 143
phenotype, 4
Philippines, 141
phone, 93
photosynthetically active radiation, xi, 34, 42, 44, 106
physiology, 4, 37, 39
Physogyra, 48
plankton, 51, 59, 140
plant, 31, 32
Plerogyra, 48
Plesiastrea, 48
Pocillopora, 48, 54
poles, 20, 22
policy, 2, 11, 12, 16, 17, 19, 94, 100, 108, 109, 111, 124, 131, 133
political, 1, 13, 17, 145
pollution, 12, 24, 40, 44, 51, 85, 86, 141
population, 3, 4, 5, 7, 14, 32, 33, 37, 38, 47, 54, 75, 78, 79, 83, 85, 86, 92, 94
Porites , 46, 47, 52, 54
portfolio, 73, 89
Portugal, iv, 17
poverty, viii, 1, 2, 3, 5, 80, 81, 92, 97, 99, 111, 116, 117, 118, 120, 127, 129, 131 151
predator, 60, 65
president, 73
price, 18, 73, 87, 90, 91, 151
pristine, 54, 105, 137, 138, 140
private, 89, 90, 118, 119, 145, 148
production, 3, 7, 8, 9, 10, 13, 15, 16, 19, 20, 21, 32, 34, 38, 54, 55, 59, 61, 64, 67, 69, 73, 85, 93, 112, 142, 144
productivity, 7, 21, 31, 39, 52, 59, 60, 67, 94, 118, 119, 129, 144
property, 11, 78, 90, 92, 119, 120, 121, 122, 123, 143, 147, 148
protein, 10, 21, 77, 93
proxies, 22, 23, 25, 27, 30, 34, 35, 36
psychology, 80, 130
Pulicit Lagoon, 77
purchasing power parity, xi, 70, 83

quota, 17, 147

rabbitfish, 61, 63, 64
radiation, xi, 6, 34, 35, 40, 42, 44, 106
rainfall, vii, 3, 22, 24, 25, 28, 29, 30, 31, 32, 36, 65, 68, 84, 85, 93, 111
rebuild, 9, 13, 121, 132, 139, 152
recovery, 9, 13, 121, 132, 139, 152
recruitment, 54, 60, 61, 62, 138, 140, 144
Red Sea, 28, 43, 44, 45, 50

redundant, 55
reinforce, 5, 97, 98, 117, 126, 131
remittances, 80
resilience, v, 3, 4, 5, 13, 14, 21, 37, 39, 41, 43, 45, 47, 49, 51, 53, 55, 73, 88, 97, 98, 99, 113, 134, 140, 140, 141, 143, 145, 149, 152, 154
resource, v, vii, 1, 2, 3, 5, 6, 7, 10, 11, 12, 13, 15, 16, 17, 19, 20, 21, 37, 62, 67, 69, 72, 73, 75, 76, 77, 78, 79, 80, 81, 85, 86, 92, 94, 97, 101, 102, 109, 110, 111, 112, 113, 114, 115, 116, 119, 120, 121, 122, 124, 124, 125, 126, 127, 128, 129, 130, 131, 133, 134, 137, 138, 139, 142, 145, 147, 149, 150, 151, 152
restriction, 65, 75, 77, 110, 114, 119, 138, 142, 143, 144, 145, 148, 151
Reunion, 30, 44, 45, 46, 50, 82, 101, 141, 151
Rift Valley, 1
risk, 4, 55, 73, 74, 76, 87, 88, 91, 92, 94, 97, 101, 106, 107, 110, 111, 115, 117, 120, 127, 137, 138
Rodrigues, 44, 45, 50, 51, 53
Roviana Lagoon, 127
roving-bandit, 92
rudderfish, 61
Russia, 19, 21
Rwanda, 1

Sahamalaza, 74, 77, 116
Sahara, 24, 69, 79, 81
Sahasoa, 103
salinity, 40, 52
saltwater intrusion, 3, 86
satellites, 25, 34, 35, 41, 52, 53, 106
sea urchin, 51, 56, 135, 137
sea wall, 4, 69, 86, 89, 129
sea-level rise, 3, 4, 67, 68, 85, 86, 90, 92, 129
sea-surface temperature, xi, 41, 106, 108
seafood, 7, 9, 10, 13, 15
seascape, 55, 140, 141, 143, 152
season, 27, 29, 30, 32, 40, 41, 73, 79, 84, 121, 127, 138, 142,
seawater, 26, 27, 43, 51, 52, 60, 66, 106
seaweed farm, 119
sediment, 22, 23, 28, 34, 40, 53, 65
sensitivity, 2, 5, 47, 48, 67, 68, 69, 70, 71, 72, 83, 105, 112, 128, 129, 133
settlement, 24, 48, 54, 57, 58, 68, 71, 86, 129
Seychelles, 17, 21, 29, 43, 44, 45, 50, 62, 63, 64, 69, 70, 71, 75, 76, 78, 79, 81, 82, 83, 95, 101, 102, 103, 104, 106, 108, 109, 110, 113, 131, 132
shock, 73, 79, 90, 94, 97, 115, 116, 128, 134
shrimp, 10, 17, 119
skeleton, 27, 40, 45, 51, 54, 57, 59
snapper, 63
snow, 24, 25
social capital, 76, 77, 102, 103, 104, 114, 122, 126
social castes, 77

social organization, 72, 75, 77, 89, 90, 116, 122, 125, 126, 152
social-ecological, 5, 6, 73, 97, 102, 110, 112, 115, 122, 134. 152, 153
Socotra, 44, 50
Solomon Islands, 119, 127
solutions, 1, 6, 93, 151
Somalia, 21, 25, 31, 50, 69, 75, 76, 78, 79, 80, 81, 82
South Africa, 50, 75, 76, 78, 79, 81, 82, 107, 117
Soviet, 17
Spain, 16, 17, 19
species, xi, 9, 13, 14, 15, 18, 20, 27, 34, 38, 39, 40, 45, 46, 47, 48, 49, 53, 54, 55, 56, 57, 58, 59, 60, 61, 62, 63, 64, 65, 66, 69, 71, 72, 77, 90, 91, 94, 105, 106, 110, 112, 127, 129, 134, 135, 136, 137, 138, 139, 140, 141, 142, 143, 144, 145, 146, 147, 148, 149, 151
spiritual, 75
sponge, 49, 51, 62
Sri Lanka, 43, 44, 45, 50
SST, xi, 23, 26, 29, 35, 41, 42, 43, 44, 1-6
stalagmite, 22, 23, 24
starvation, 1
statistic, 8, 17, 48, 70, 71, 73 78
stewardship, 11, 13, 92, 147, 148
storm, 3, 31, 53, 59, 68, 81, 84, 89, 90, 91, 92, 129
stress, 3, 4, 5, 27, 28, 32, 34, 36, 37, 38, 40, 41, 44, 45, 46, 51, 53, 54, 55, 56, 68, 75, 79, 99, 100, 107, 112
subsidies, 13, 15, 74, 88, 115
subsidize, 17, 75
subtractability, 11
success, viii, 1, 4, 12, 38, 54, 58, 61, 69, 84, 85, 88, 100, 109, 113, 118, 119, 121, 122, 124, 126, 127, 141, 142, 147, 148, 149, 152
succession, 38, 61
Sudan, 1
suffer, 2, 3, 44, 45, 47, 49, 62, 72, 100, 112
Sulawesi, 119
Sumatra, 23, 26, 27
supernatural, 80
surgeonfish, 61, 105
surimi, 9
susceptible, 34, 45, 46, 47, 49, 51, 59, 62, 65, 67, 72, 101, 105, 106, 119, 112, 140, 145, 146
sustainable, xi, 1, 9, 10, 11, 12, 13, 14, 17, 112, 116, 119, 133, 137
Swedish International Development Corporation Agency, viii, xi
switch-point, 135
Symbiodinium, 38, 39, 40
synergy, 104

taboo, 75, 77, 131, 132, 143
Takaungu, 74
Tanzania, 27, 44, 46, 49, 50, 54, 69, 70, 71, 74, 75, 76, 78, 79, 81, 82, 83, 86, 87, 95, 97, 98, 99, 101.102, 104, 108, 109, 111, 113, 130, 132, 142

tax, 17, 39, 46, 47, 48, 49, 53, 54, 80, 88, 105, 139, 140
technology, 93, 102, 103, 131, 147, 150
teleconnection, 31
temperature, xi, 2, 3, 20, 22, 23, 24, 25, 26, 27, 28, 29, 30, 31, 34, 35, 36, 37, 40, 41, 42, 43, 44, 45, 46, 48, 49, 51, 52, 53, 54, 57, 59, 60, 66, 68, 90, 91, 93, 105, 108
tengefu, 120
territorial, 11, 16
Thailand, 19, 50
threshold, 1, 5, 6, 104, 119, 117, 118, 127, 134, 135, 136, 137, 138, 148
Tibetean Plateau, 24, 25, 96
time, 1, 2, 4, 17, 19, 26, 27, 28, 29, 30, 34, 36, 37, 38, 39, 41, 47, 53, 55, 56, 58, 61, 62, 65, 76, 77, 81, 84, 87, 91, 92, 92, 94, 109, 110, 119, 120, 121, 122, 123, 125, 127, 129, 130, 132, 138, 142, 143, 149, 151
time lag, 61, 62, 65
Tobago, 122
topographic complexity, 57, 58, 59, 62
tourism, 55, 69, 70, 85, 86, 87, 88, 92, 109, 110, 111, 112, 139, 141, 152
trade-off, 38, 49, 67, 140
tragedy of the commons, 11
transparency, 16, 17, 18, 115, 122, 12, 126, 150
transportation, 69, 86, 92
trawl, 13
Trinidad, 122
tropic, vii, xi, 10, 18, 20, 21, 22, 23, 27, 29, 30, 31, 32, 30, 41, 43, 52, 53, 56, 59, 60, 65, 68, 147
tsunami, 59
tuna, 10, 17, 18, 91
turf, 49, 50, 60, 61

ultraviolet, xi, 6, 34, 35 40, 42, 44, 106
United Kingdom, UK, 10, 19, 25,
United Nations Convention on the Law of the Sea, xi, 11
United States, iv, 9, 12, 13, 14, 19, 21, 25, 91
unreported, 9, 16, 18
upwelling, 25, 26, 27, 31, 32, 65
Utange, 74, 103

value, 7, 9, 10, 15, 17, 18, 19, 29, 50, 54, 55, 64, 65, 70, 86, 87, 89, 90, 91, 105, 106, 107, 109, 110, 111, 113, 116, 135, 138
value chain, 135
vanilla, 73
Vanuatu, 143
Velondriake, 121
Vietnam, iv, 19, 122
volatile, 91

vulnerable, v, 3, 4, 5, 10, 11, 35, 43, 44, 48, 49, 55, 56, 67, 68, 69, 71, 73, 75, 76, 77, 78, , 79, 81, 83, 84, 85, 86, 87, 100, 112, 122, 128, 130, 133, 141, 151
Vuma, 74

warm, 1, 2, 3, 20, 22, 23, 24, 25, 26, 27, 29, 30, 31, 32, 32, 36, 41, 44, 45, 46, 51, 52, 53, 54, 59, 84, 105
warn, 69, 81, 84, 99, 90, 91, 93, 111, 115, 123, 127, 137, 152
water flow,
wealth, 6, 7, 8, 12, 13, 96, 97, 103, 117, 118, 122, 129, 130, 131, 134, 147
wedding, 117
weighting, 103, 105, 112
well-being, vii, 36, 67, 83, 117, 118, 122, 149
western Indian Ocean, xi, 1, 2, 3, 43, 46, 48, 49, 53, 54, 57, 67, 68, 70, 73, 75, 78, 79, 80, 81, 82, 83, 84, 87, 94, 100, 101, 102, 107, 108, 109, 117, 121, 122, 123, 128, 130, 132, 134, 135, 137, 138, 139, 140, 141, 147, 149
Western Indian Ocean Marine Science Association, ix, xi
wild, xi, 8, 9, 10, 12, 85, 121, 141
wind, 28, 31, 42, 45, 91, 106, 108
WIOMSA, ix, xi
World Trade Organization, 15
World Wildlife Fund, xi, 12
wrasse, 63, 105, 143

Zambezi, 86
Zanzibar, 41, 50, 87
Zimbabwe, 1, 127, 130
zooxanthellae, 39, 46